Springer Theses

Recognizing Outstanding Ph.D. Research

Aims and Scope

The series "Springer Theses" brings together a selection of the very best Ph.D. theses from around the world and across the physical sciences. Nominated and endorsed by two recognized specialists, each published volume has been selected for its scientific excellence and the high impact of its contents for the pertinent field of research. For greater accessibility to non-specialists, the published versions include an extended introduction, as well as a foreword by the student's supervisor explaining the special relevance of the work for the field. As a whole, the series will provide a valuable resource both for newcomers to the research fields described, and for other scientists seeking detailed background information on special questions. Finally, it provides an accredited documentation of the valuable contributions made by today's younger generation of scientists.

Theses are accepted into the series by invited nomination only and must fulfill all of the following criteria

- They must be written in good English.
- The topic should fall within the confines of Chemistry, Physics, Earth Sciences, Engineering and related interdisciplinary fields such as Materials, Nanoscience, Chemical Engineering, Complex Systems and Biophysics.
- The work reported in the thesis must represent a significant scientific advance.
- If the thesis includes previously published material, permission to reproduce this must be gained from the respective copyright holder.
- They must have been examined and passed during the 12 months prior to nomination.
- Each thesis should include a foreword by the supervisor outlining the significance of its content.
- The theses should have a clearly defined structure including an introduction accessible to scientists not expert in that particular field.

More information about this series at http://www.springer.com/series/8790

David Carreto Fidalgo

Revealing the Most Energetic Light from Pulsars and Their Nebulae

Doctoral Thesis accepted by
the Complutense University of Madrid, Madrid,
Spain

 Springer

Author
Dr. David Carreto Fidalgo
Faculty of Physics
Complutense University of Madrid
Madrid, Spain

Supervisor
Prof. Marcos López Moya
Faculty of Physics
Complutense University of Madrid
Madrid, Spain

ISSN 2190-5053   ISSN 2190-5061   (electronic)
Springer Theses
ISBN 978-3-030-24196-4   ISBN 978-3-030-24194-0   (eBook)
https://doi.org/10.1007/978-3-030-24194-0

This Springer imprint is published by the registered company Springer Nature Switzerland AG
The registered company address is: Gewerbestrasse 11, 6330 Cham, Switzerland

Para António.

Supervisor's Foreword

Very-high-energy (VHE) gamma-ray astronomy provides a unique view to the most extreme phenomena in the Universe. The observation of such energetic gamma-ray radiation implies the presence in the source of some kind of mechanism which is able to accelerate charged particles to ultra-relativistic energies. Pulsars, and their surrounding nebulae, are one of the main classes of galactic sources where these processes take place.

Pulsars are fast-rotating and highly magnetized neutron stars. Although they are mostly visible in radio, more than two hundred have been detected by the *Fermi*-LAT space telescope. Above a few GeV, the emission from *Fermi*-LAT pulsars typically drops exponentially, as predicted by classical pulsar models. Therefore, not long before Dr. David Carreto Fidalgo started his thesis, it was thought that pulsars could not emit in the VHE energy domain. The situation changed when the MAGIC telescopes, located in the Canary Island of La Palma, detected emission from the Crab Pulsar.

David's main goal was to study the high-energy radiation detected in the Crab Pulsar and to answer the long-standing question up to what energies pulsars are able to radiate. To this end, he analyzed more than 400 h of the Crab Pulsar data taken by MAGIC. The analysis of such unprecedented large data set, collected over nearly a decade of observations, was a real challenge. The arduous work that the author presents in a great detail resulted in a great discovery: Pulsars can emit radiation beyond TeV energies. The implications of pulsed TeV gamma rays for pulsar models are clearly discussed in the thesis. The results suggest that the emission at these energies is produced in the outer regions of pulsar magnetospheres, via inverse Compton scattering.

The detection of ultra-energetic pulses from the Crab provided a unique data set to address questions of fundamental physics. Quantum gravity models predict the violation of the Lorentz invariance, that is, the change of the speed of light with energy. If so, photons emitted in the Crab Pulsar at the same time but with different energies would be detected with a relative delay. Exploring a new methodology, based on Bayesian inference, the author is able to obtain competitive limits on the invariant energy scales of the speed of light.

To shed more light on the pulsar emission mechanism at the highest energies detected in the Crab, we would need to find other pulsars exhibiting a similar behavior. The thesis addresses also this topic, by searching among the *Fermi*-LAT pulsars for candidates to emit in the VHE range. The author found a millisecond pulsar that would be a prime target for gamma-ray observatories in the Southern Hemisphere, such as the upcoming CTA-South.

The final part of the thesis deals with the search for pulsar wind nebulae. Although they are a common class of VHE gamma-ray emitters in the inner Milky Way, they are hardly found in the outer part of our galaxy. This motivated the author of this thesis to conduct observations around a young and energetic pulsar, PSR J0631+1036, situated near the galactic anticenter. The results corroborate the relationship between the luminosity of pulsar wind nebulae and the interstellar radiation field.

Apart from the remarkable detection of the most energetic light ever been seen from a neutron star, the reader will find in the thesis a comprehensive introduction to the research field of gamma-ray astronomy, covering both its experimental and theoretical aspects. And all this is written in a pleasant and didactic style that the reader will surely enjoy.

Madrid, Spain
May 2019

Prof. Marcos López Moya

Preface

The observation of very-high-energy (VHE, >100 GeV) gamma rays is key in studying the nonthermal sources of radiation in our Universe. Pulsars and pulsar wind nebulae (PWNe) are two source classes that are known to emit VHE gamma rays. While pulsar wind nebulae are the dominant VHE gamma-ray source class in our galaxy, only two pulsars have been detected above 100 GeV so far. Most pulsar models explain gamma-ray emission via synchro-curvature radiation in the radiation-reaction-limited regime, which leads to a sharp cutoff in the pulsar spectrum at energies of a few GeV. However, the detection of pulsed emission from the Crab Pulsar up to hundreds of GeV by MAGIC and VERITAS suggests that classical pulsar models do not provide a full picture of the emission mechanisms at work. TeV pulsar wind nebulae, on the other hand, are observed via their inverse Compton radiation and are primarily found around young and energetic pulsars located toward the inner Milky Way. Detections of TeV PWNe in the outer part of our galaxy are scarce but could provide valuable input for the connection between the interstellar radiation field and the PWN luminosity.

The general motivation for studying *pulsars* and *pulsar wind nebulae* at very high energy (VHE, >100 GeV) is manifold. Pulsars are exotic objects with core densities several times higher than atomic nuclei and surface gravities 10^{11} times stronger than the Earth's. At the same time, they spin up to a factor 10^7 faster than the Earth and their magnetic fields can surpass the Earth's by 15 orders of magnitude. Because of these extreme conditions, pulsars can give insights into physics of ultra-dense matter, tests for general relativity, and the promise of a direct detection of low-frequency gravitational waves. Most of them use their enormous rotational energy to emit electromagnetic radiation via detailed processes we do not fully understand yet. The modeling of their emission drives ever-more sophisticated electrodynamic calculations, for which VHE observations provide valuable input.

Most of the rotational energy of a pulsar is dissipated via relativistic winds that interact with the surrounding medium to generate luminous pulsar wind nebulae. These sources allow us to probe relativistic shocks, particle acceleration as well as

particle diffusion and propagation in our galaxy. Pulsar wind nebulae are prime
candidates to explain the puzzling positron excess above 10 GeV in the cosmic ray
flux. To solve this puzzle, VHE energy observations of these objects are key.

The principal goal of this thesis is to study the very-high-energy emission of the
Crab Pulsar. We aim to answer the long-standing question up to what energies
pulsars are able to radiate and what is the emission mechanism behind it. We further
exploit the pulsed VHE emission from the Crab to investigate fundamental physics
testing for Lorentz invariance violation (LIV), in terms of a wavelength-dependent
speed of light. To deepen our understanding of the pulsar emission mechanism at
the highest energies, further pulsars have to be discovered above 100 GeV. To this
end, we search among pulsars already detected at around 1 GeV for the best
candidates to emit gamma rays in the VHE range. Another aim of this thesis is to
discover a new pulsar wind nebula toward the outer part of our galaxy with the
MAGIC telescopes.

The thesis is structured into three parts:

Part I
This introductory part gives a brief overview of the instruments and astrophysical
sources in VHE astrophysics. It also introduces the reader to pulsar and pulsar wind
nebulae physics and describes in detail the MAGIC experiment together with the
Imaging Atmospheric Cherenkov Telescope (IACT) technique.

Part II
The main part of this thesis starts with the discovery of VHE emission from the Crab
Pulsar above 400 GeV. We characterize its pulse profile as well as the
phase-resolved VHE spectra up to ~ 1.2 TeV and discuss our results with respect to
different emission scenarios. In Chap. 6, we exploit the VHE emission of the Crab
Pulsar to test for Lorentz invariance violation, in terms of a wavelength-dependent
speed of light. The last chapter of this part is devoted to the search for VHE pulsar
candidates with *Fermi*-LAT.

Part III
We describe the search of a TeV pulsar wind nebula around the young and energetic gamma-ray pulsar PSR J0631+1036 with the magic telescopes. PSR J0631 lies
near the galactic anticenter, and a tempting hot spot at its position was reported by
the Milagro collaboration based on their full eight-year data set.

Summary and Conclusions
In the last chapter of this thesis, we give an extended summary and concluding
remarks about the results in view of the future Cherenkov Telescope Array (CTA)
Observatory.

Appendix
The appendix provides more background information about pulsar timing and a
quick overview of physical interaction processes in VHE astrophysics. We also
describe the On-Site Analysis (OSA) chain of MAGIC, which is the system in charge

of the low-level analysis of MAGIC data. During his Ph.D. studies, the author worked continuously on OSA to improve and maintain its workflow. In the end, we provide some more technical details about the main analysis of this thesis.

Madrid, Spain   Dr. David Carreto Fidalgo

Acknowledgements

Crafting this thesis was an effort of many hands, and there are many people whom it is a pleasure to thank for their help, support, and advice during the course of my Ph.D. studies.

First of all, I want to thank María Victoria for giving me the opportunity to work in the Grupo de Altas Energías and to conduct my Ph.D. thesis at the Universidad Complutense de Madrid. I am also grateful for the financial support by the Spanish Ministerio de Economía, Industria y Competitividad.

I would like to express my special appreciation and warmest thanks to my supervisor and mentor Marcos. Without your advice, direction, and encouragement, this work would only have been a shadow of itself. I am also most grateful for working with José Luis who, apart from the financial support in the end, always had inspiring thoughts and encouraging words for me. With great pleasure, I have worked with former and present members of the fabulous age group and I want to thank them all for their daily chats, discussions, and jokes: Juan Abel, Fernando, Jaime, Konstancja, Ester, Luis Angel, Diego, Pablo, Dani, Alberto, Daniel, Irene, John, Lab, and Valeria. Of course, I want to thank my brothers and sisters who walked with me through the valley of death: Simon, Tarek, and Mireia; it was great fun with you guys! I am also very grateful for the experiences inside MAGIC, a wonderful collaboration with wonderful people and an impressive experiment. I learned a lot within the collaboration, and my research would not have been possible without the help and support from so many *magical* members. Special thanks to the people who endured an observational shift with me and to all the current and former Ph.D. students from the IFAE and MPI group.

I want to express my sincere gratitude to Pablo for giving me the opportunity to spend a research stay at the HKU Laboratory for Space Research. He and the whole group welcomed me with open arms and took great care of me right from the start. So many thanks to you: Abdi, Xuan, Yong, Xia, Quentin, David, and Ivan. I am also much obliged to Prof. K. S. Cheng and Dr. Stephen Ng for their support during the stay.

Last but not least, I want to thank my much better half. This thesis would definitely not have seen the light of day if it were not for your daily and selfless support, especially during the last stage of this adventure. So at least half the credit goes to you. And little man; I am not sure if I should thank you for all the sleepless nights and the whole new level of tiredness you taught me. But I guess you helped me to keep seeing things in the right perspective, so thank you for that.

Contents

Part I
The Very-High-Energy Sky and the MAGIC Telescopes

Chapter 1
An Introduction to Very-High-Energy Astrophysics

The aim of the astrophysical sciences is two-fold - the application of the laws of physics in the extreme physical conditions encountered in astronomical systems, and the discovery of new laws of physics from observation.

Malcolm S. Longair, 2010

Astronomy is one of the oldest natural sciences and has always been an important part of human culture. Astronomical observations can be traced back to the Stone Age manifesting itself in ancient beliefs and structures such as references to Egyptians gods or Stonehenge. For most part of astronomy's history observations were limited to the light visible to the human eye. Only in the 20th century technological advances, the development of new techniques and serendipities lead to the opening up of the whole of the electromagnetic spectrum. At the low-energy end of the spectrum, the first observations were made from the ground by radio telescopes in the late 1930s, whereas at the high-energy end the first evidence of cosmic gamma rays came from detectors aboard satellites in the late 1960s. The reason why high-energy astrophysics is primarily conducted from space, lies in the Earth's atmosphere as illustrated by Fig. 1.1. While for the visible and the radio part of the spectrum the atmosphere is mostly transparent, X-rays and gamma rays are completely absorbed by it. Only at the highest energy end, above a few tens of GeV, observing the violent absorption process in the atmosphere makes the detection of gamma rays possible from the ground.

In practice the energy spectrum of cosmic gamma rays is roughly divided into three ranges following loosely the detection technique: *soft* ($\lesssim 10\,\text{MeV}$) and *high-energy* (HE, $\lesssim 50\,\text{GeV}$) gamma rays are recorded directly by instruments on high altitude balloons or aboard satellites, while *very-high-energy* (VHE, $\gtrsim 100\,\text{GeV}$) gamma rays are usually observed indirectly by telescopes on the ground. Our atmosphere not only protects us from harmful electromagnetic radiation, but also from very energetic particles with intrinsic mass known as cosmic rays. Their discovery in 1912 is

© Springer Nature Switzerland AG 2019
D. Carreto Fidalgo, *Revealing the Most Energetic Light from Pulsars and Their Nebulae*, Springer Theses, https://doi.org/10.1007/978-3-030-24194-0_1

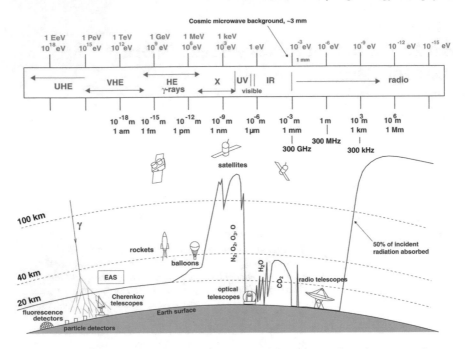

Fig. 1.1 Illustration of the electromagnetic spectrum and its corresponding telescopes or observations techniques. The visible part of the light constitute only a tiny fraction of the total electromagnetic spectrum from radio to infrared (IR) and ultraviolet (UV), to X-rays and gamma rays. While high-energy (HE) gamma rays are detected by instruments on high altitude balloons or satellites, very-high-energy (VHE) and (theoretically) ultra-high-energy (UHE) gamma rays are indirectly observed from ground via the absorption process in the atmosphere producing a so-called Extensive Air Shower (EAS, see Sect. 3.1). Figure taken from López Coto [1]

attributed to Victor Hess who observed an increasing ionization rate while ascending to an altitude of $\sim$5 km in a free balloon flight. Cosmic rays consist almost entirely of atomic nuclei ($\sim$98%) out of which $\sim$87% are simple protons [2]. Their origin is still not fully understood and finding the sources of the most energetic cosmic rays with energies exceeding $\sim 10^{15}$ eV is a very active research area still today. The main challenge is that cosmic rays loose the information about their origin on their way to earth because of deflections by intergalactic, galactic, solar and planetary magnetic fields. Photons on the other hand propagate nearly undeflected through the universe, and therefore accurately pinpoint their sources in space. Since physical processes involving very energetic leptons or hadrons (cosmic rays) often lead to the production of VHE gamma rays, they can be used to locate the sites of cosmic ray production and to reveal acceleration mechanisms together with the physical environment at the source (see Appendix A for details). At the same time, the experimental detection techniques of the most energetic cosmic rays and VHE gamma rays are both based on the observation of the absorption processes in the atmosphere. Hence, VHE astrophysics grew hand in hand with cosmic ray physics and their joint field of study is often referred to as astroparticle physics, which also includes other research areas related to particle physics, such as neutrino astronomy or direct dark matter searches.

In the first section of this chapter we will introduce current VHE gamma-ray detectors and mention future next generation observatories. The second section will summarize known source types of cosmic gamma rays and conclude the chapter.

1.1 Instruments for TeV Astronomy

The direct detection of cosmic gamma rays is only possible from space or high altitude balloons as illustrated by Fig. 1.1. However, the very low photon fluxes in the VHE regime (a few gamma rays per square meter per year above 1 TeV for strong sources [3]) makes space-based instruments little effective due to the detector's limited collection area. While the effective detector area of the *Fermi* Large Area Telescope (LAT), which is currently the largest gamma-ray telescope in space, is around $\sim$1 m^2 at 1 TeV [4], Imaging Atmospheric Cherenkov Telescopes (IACTs) on the ground reach areas of $\sim$10^5 m^2 (see Chap. 3). These huge areas are achieved by effectively exploiting the Earth's atmosphere as part of the instrument.

For VHE gamma rays the mass absorption coefficient is dominated by the electron-positron pair production in the Coulomb field of a nucleus. In the same high energy regime, the dominant interaction for electrons and positrons with matter is Bremsstrahlung via acceleration in nuclear Coulomb fields [5]. The interplay between the two processes leads to an electromagnetic cascade and as long as the next-generation photons, emitted via Bremsstrahlung by the secondary electron-positron pairs, remain above the pair-production threshold, the particle shower continuous to grow exponentially. Due to the relativistic speeds of the electrons and positrons, the Bremsstrahlung is beamed in a narrow cone about the particle's trajectory resulting in a well-contained shower of particles that allow inference about the properties of the incident photon (see Sect. 3.1 for details). All VHE gamma-ray telescopes are based on recording this particle shower to reconstruct the energy and the incoming direction of the primary gamma ray. While space-born gamma-ray telescopes need to promote the pair conversion by themselves, ground-based telescopes use the Earth's atmosphere as conversion medium resulting in huge effective detector areas as mentioned above. A major drawback of using the atmosphere as part of the detector is the missing capability of artificially providing a calibration source and therefore having to rely solely on computer simulations and astronomical standard candles, such as the Crab nebula.

In the following we will provide a quick overview of the current instruments used in VHE astronomy starting with the space-born telescope, the *Fermi* LAT (see Fig. 1.2). On the ground we will divide the telescopes into detectors that directly measure the shower particles reaching the ground, so-called *particle sampler*, and

Fig. 1.2 Images of some of the instruments used in the very-high-energy range. An artistic impression of the *Fermi* satellite with its Large Area Telescope (LAT) is shown on the top, left and right are pictures of the HAWC observatory and the MAGIC telescopes, respectively (see text). Images taken from the web pages of the *Fermi*, HAWC and MAGIC collaborations. Background image taken from http://www.imagico.de/pov/earth_atmosphere.php, last accessed 17/11/2017.

detectors that image the particle shower via the emitted Cherenkov light, known as Imaging Atmospheric Cherenkov Telescopes (IACTs, see Chap. 3).

1.1.1 *Fermi*-LAT

Launched in 2008, the *Fermi* Gamma-ray Space Telescope orbits the earth at an altitude of around 2000 km and is mostly intended to perform an all-sky survey. It hosts two instruments: the Large Area Telescope (LAT) and the Gamma-ray Burst Monitor (GBM). The latter will not be discussed in this work. The *Fermi* LAT can be thought of as a sequel to the Energetic Gamma Ray Experiment Telescope (EGRET) aboard NASA's Compton Gamma Ray Observatory satellite that had been operating for 9 years when it was deliberately de-orbited in 2000 [6]. *Fermi*-LAT covers the energy range from ∼20 MeV to more than 300 GeV and has detected over 3000 point sources so far [7, 8]. The instrument consists of three detector subsystems [9]: (i) a tracker/converter (TKR) that promotes pair conversion and measures the direction of the incidence particle; (ii) a calorimeter (CAL) that provides energy measurements as well as some imaging capabilities; (ii) an Anticoincidence Detector (ACD) that rejects cosmic rays.

The TKR comprises 16 planes of tungsten foils in which energetic photons can convert to electron-positron pairs. These foils are interleaved with layers of paired x-y

Silicon Strip Detector (SSD) planes that record the passage of charged particles, thus measuring the tracks of the secondary electron-positron pairs. The first 12 converter foils have a thickness of ~3% of a radiation length of tungsten, minimizing the separation of the converter foils from the following SSD planes and hence minimizing the effects of multiple scattering. This results in a better determination of the incident particle's direction. The last four tungsten foils right above the CAL module are ~6 times thicker to ensure a large conversion probability. The former part is referred to as front section, the latter part is called back section. The CAL is an electromagnetic calorimeter consisting of 96 CsI crystal logs that serve both to continue to seed the electromagnetic cascade and to scintillate as the main energy loss mechanism turns from pair production to ionization of the CsI atoms. Two photodiodes at each end of the logs provide two readout channels to cover the large dynamic range of energy deposition in the crystal from 2 MeV to 70 Gev. The total vertical depth of the calorimeter is 8.6 radiation lengths, for a total instrument depth of 10.1 radiation lengths. The background of charged particles (cosmic rays) can exceed the flux of the desired gamma-ray signal by up to 10^5 [5]. Cosmic electrons and positrons are especially problematic since they generate particle showers in the TKR and CAL very similar to those initiated by a gamma ray. For this reason the ACD tries to detect these particles before entering the TKR/CAL and reject the corresponding triggers. It consists of scintillating plastic tiles covering the top and the sides of the instrument.

Regarding the performance of *Fermi*-LAT,[1] its differential sensitivity in comparison to other VHE instruments is shown in Fig. 1.3. Its total angular resolution (68% containment) is highly energy dependent and goes from ~10° at the lowest energies to ~0.1° above ~10 GeV. The energy resolution is around ~5% at 10 GeV and increases to ~25% towards the ends of the instrument's energy coverage.

1.1.2 Imaging Atmospheric Cherenkov Telescopes

While the sensitivity of *Fermi*-LAT decreases rapidly above 10 GeV due to the limited size of the detector, ground-based observations are able to efficiently detect gamma rays above ~50 GeV by recording the particle showers induced in the atmosphere. However, one of the main challenges faced by gamma-ray astronomy from the ground, is the background rejection since the nature of a particle shower, that is if it was induced by a gamma ray or a cosmic ray, can only be determined after its recording and only up to a certain degree of certainty depending on the shower size. Thus, the vast majority of particle showers recorded from ground belong to cosmic rays. In the IACT technique atmospheric showers are detected via their emitted Cherenkov light and hence the technique is able to record showers that partially or completely died before reaching the ground. This has the advantage that most of the shower's energy will be deposited in the atmosphere, effectively using it as a

---

[1]For details, see http://www.slac.stanford.edu/exp/glast/groups/canda/lat_Performance.htm , last accessed 31/10/2017.

Table 1.1 Key parameters of the current three major IACTs

Parameter	H.E.S.S. I/II	VERITAS	MAGIC
Site	Gamsberg, Namibia	Arizona, USA	La Palma, Spain
Altitude (m)	1800	1300	2200
Nr. of telescopes	4/1	4	2
Layout area (m^2)	428	424	234
Dish size (m)	12/28	12	17
Field of view (°)	5/3.2	3.5	3.5
Pixels per camera	960/2048	499	1039
Angular resolution (°)	$\gtrsim$0.05	$\gtrsim$0.06	$\gtrsim$0.06
Energy threshold (GeV)	~100/ < 50	~100	~70

Notes The parameters for the H.E.S.S. array are given with respect to the 4 small telescopes, H.E.S.S. I, and the latest bigger telescope, H.E.S.S. II (see text). The second row specifies the approximate altitude in meters above sea level (a.s.l.). The dish size is given in meters for the diameter of an equivalent, in terms of area, circular dish. The definition of the energy threshold in case of IACTs is given in Sect. 3.4. Values adopted from [3, 10–13]

calorimeter. We will extensively discuss air showers and in particular the imaging technique in Chap. 3. Current major IACTs are the H.E.S.S.[2] telescopes near the Gamsberg mountain, Namibia, the VERITAS[3] in southern Arizona, USA, and the MAGIC[4] telescopes at the Roque de Los Muchachos Observatory in La Palma, Canary Islands (Spain). The phase 1 of the H.E.S.S. project consisted of 4 telescopes with a dish size of 12 m in diameter that became fully operational in 2003. A much larger fifth telescope with a dish size of 28 m, called H.E.S.S. II, joined the system in 2012. Located in the Southern Hemisphere, H.E.S.S. has a prime spot for observing the inner part of our galaxy and scanning the Galactic Plane. VERITAS also relies on four 12 m telescopes situated in the Northern Hemisphere, which came online in 2007. In 2009 and 2012 the collaboration went for a slight change in the array layout and an upgrade of the cameras, respectively, to improve the instrument's sensitivity. The MAGIC telescopes will be introduced and discussed extensively in Sect. 3.3. A comparison of the key parameters of the three major IACTs is displayed in Table 1.1. The future next generation IACT project is the Cherenkov Telescope Array (CTA). It will consist of two observatories, one in the northern and one in the Southern Hemisphere and is planned to come online in 2021.[5]

[2]High Energy Stereoscopic System, https://www.mpi-hd.mpg.de.
[3]Very Energetic Radiation Imaging Telescope Array System, https://veritas.sao.arizona.edu.
[4]Major Atmospheric Gamma Imaging Cherenkov, https://magic.mpp.mpg.de/.
[5]https://www.cta-observatory.org/project/status, last accessed 02/11/2017.

1.1.3 Particle Sampler

Particle sampler are able to directly detect the particles of the air shower tail reaching the ground and hence provide an exact snapshot of the shower at the moment it hits the ground. Since only the most energetic showers are able to reach the ground, the energy threshold of these telescopes is rather high depending on the altitude of the observatory. The particles of the shower can be detected via either scintillation counters (used by the Tibet Air Shower Array, [14]), resistive plate chambers (used by the Astrophysical Radiation with Ground-based Observatory at Yang Ba Jing, ARGO- YBJ, [15]) or water Cherenkov detectors. The latter has proven to be the most sensitive method and was successfully explored by the Milagro experiment between 1999 and 2008 [16]. Milagro was composed of an opaque 60×80 m^2 water pond surrounded by a 200×200 m^2 array of 175 smaller water tanks. The central pond and the water tanks were equipped with photomultiplier tubes (PMTs) that measured the Cherenkov radiation produced by the relativistic shower particles entering the water. Its successor, the High Altitude Water Cherenkov (HAWC) detector, was inaugurated in 2015 reusing most of Milagro's PMTs and front end electronics. In contrast to Milagro, HAWC is not using a big central pond but relies on 300 identical opaque water tanks with a diameter of $\sim$7 m arranged in a compact layout resulting in a detector area of about 22000 m^2. The HAWC observatory is located in the Sierra Negra in Mexico at an altitude of 4100 m, significantly higher than the 2650 m of the Milagro detector. The increased altitude leaves the detector closer to the shower maximum providing on average six times the number of shower particles and decreasing the energy threshold to about $\sim$1 TeV depending on the source spectrum and its declination [17]. In the sampling technique, the direction of the primary gamma ray is reconstructed using the signal arrival times in each detector and the spatial distribution of the shower particles on the ground. The angular resolution of HAWC ranges from $\sim$0.2° to $\sim$1.0°. Since particle sampler only have access to the tails of the air shower, unlike the imaging technique of IACTs, their discrimination power between gamma-ray and cosmic-ray showers is usually worse at lower energies compared to IACTs. At the same time, providing only a snapshot of the shower, their energy resolution is approximately $\gtrsim$50% in the whole energy range greatly complicating the precise measurement of a source spectrum.[6] On the other hand particle sampler have the advantage of a wide field of view ($>$1.5 sr) and a duty cycle close to 100% [17].

Figure 1.3 compiles the differential sensitivity curves of the current major instruments for VHE gamma-ray astronomy, and compares them to the two future CTA observatories in the Northern and Southern Hemisphere. Below $\sim$50 GeV observations usually rely on *Fermi*-LAT, whereas the range between $\sim$50 GeV and $\sim$10 TeV is dominated by current IACTs. Above $\sim$10 TeV the HAWC observatory is currently the most sensitive instrument owing to its large detector area and duty cycle. Due to their large sky coverage and the overlapping energy ranges, *Fermi*-LAT and HAWC are

---

[6]https://www.hawc-observatory.org/.

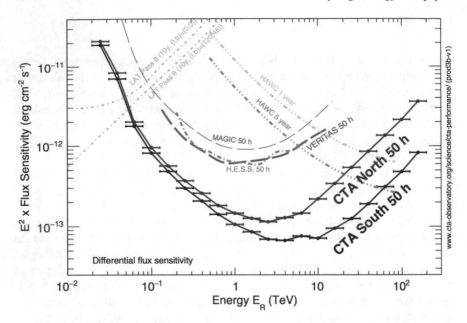

Fig. 1.3 Differential sensitivity curves for current VHE gamma-ray telescopes and the future CTA observatories. The curves correspond to 5 standard deviation detections of a point source in the indicated amount of time. The comparison of the different instruments in only indicative, as the method of calculating the sensitivity and the criteria applied are different. Figure taken from The Cherenkov Telescope Array Consortium, et al. [13]

well suited to provide observational guidance to IACTs that typically feature narrow field of views of about ∼4°. Especially the *Fermi* LAT is often used to extrapolate source fluxes to the TeV regime and predict detection probabilities for IACTs. Recently the *Fermi*-LAT collaboration released a catalog of 360 high-energy sources firmly detected at energies above 50 GeV [18].

Apart from the future CTA observatory, next generation VHE gamma-ray detectors include the ambitious Large High Altitude Air Shower Observatory (LHAASO, [19]) and the Hundred Square-km Cosmic Origin Explorer (HiSCORE, [20]), which will be able to efficiently detect gamma rays above ∼10 TeV, sometimes also dubbed ultra-high-energy (UHE) gamma rays. For a review of future cosmic gamma-ray detectors, ground-based as well as space-born, the reader is referred to [21].

1.2   Sources of Very-High-Energy Gamma Rays

Most of the visible light reaching the Earth from outer space is of thermal nature produced in hot objects such as stars or accretion disks. Following Wien's displacement law a black-body radiation with its maximum at around 1 MeV would correspond

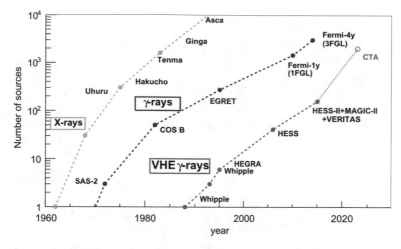

Fig. 1.4 Evolution of number of detected sources with time in different energy ranges, also known as the *Kifune plot*. The rise of VHE sources (red curve) is comparable to the one in the X-ray band (green curve) and in the gamma-ray band accessible from space (blue curve). The labels indicate the instruments responsible for the rise. For the CTA observatory the prediction is marked as an empty circle. Figure taken from de Naurois and Mazin [10] (Color figure online)

to a body temperature of $\sim 10^9$ K, exceeding by several orders of magnitude typical accretion disk temperatures of $\sim 10^{5-6}$ K around quasars [22]. The most energetic thermal continuum emission is usually observed in the X-ray band at a few keV produced by thermal bremsstrahlung. Hence, cosmic gamma rays are thought to originate from non-thermal emission processes, such as synchrotron emission, inverse Compton scattering or neutral pion decay (see Appendix A). The window to this non-thermal universe has only been pushed wide open during the last decade with the launch of the *Fermi* satellite and the success of the third generation IACTs (mainly H.E.S.S., VERITAS and MAGIC). While more than 3000 sources have been discovered by *Fermi*-LAT above 100 MeV [8], ground-based IACTs and particle sampler have so far detected VHE emission from about 200 sources (number estimated from the TeVCat[7]). In both energy ranges the number of sources as a function of time shows an almost exponential rise, a behavior that is often observed after the opening of a new window to the electromagnetic spectrum as shown in Fig. 1.4. For the time being, the number of VHE sources seems not to be limited by their frequency but by the sensitivity of the instruments, which is affirmed by simulations for the future CTA observatory predicting an approximately tenfold increase of VHE sources.

Figure 1.5 shows the distribution of VHE gamma-ray emitters in the sky color-coded by their source type. In the following we will provide a short list of known VHE gamma-ray source types dividing them between galactic and extragalactic origin. Specifications about the number of sources detected for each type are taken from the TeVCat.

---

[7]The TeVCat is an online catalog for TeV Astronomy first presented by [23]. An up to date version can be found under http://tevcat.uchicago.edu.

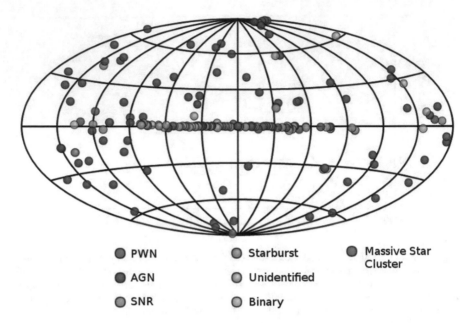

Fig. 1.5 Distribution of VHE energy sources in the sky using Galactic coordinates. The galactic plane clearly sticks out by its source density and variety of source types. Around 26% of the source are still of unidentified nature, mostly lacking associations in other energy bands. Figure reproduced from the TeVCat, http://tevcat.uchicago.edu, last accessed 10/11/2017 (Color figure online)

1.2.1 Galactic Accelerators

Pulsars and pulsar wind nebulae   Pulsars are highly magnetized neutron stars that spin at a frequency of about $0.1 - 700$ Hz and emit beams of electromagnetic radiation from radio to VHE energies. They are able to illuminate their surroundings by injecting relativistic particles and can power a Pulsar Wind Nebula (PWN). This system will be discussed in detail in the next chapter.

Supernova remnants   A supernova explosion of a massive star reaching the final stage of its stellar evolution, leaves behind an expanding nebula of ejected stellar material forming a Supernova Remnant (SNR, [24]). During the expansion the nebula sweeps up the interstellar material building shock fronts, where particles can be accelerated via Diffusive Shock Acceleration (DSA, [25]). For this reason, and because of the enormous mechanical energy output of a supernova explosion ($\sim 10^{44}$ J), SNRs are the primary candidates for sources of high-energy cosmic rays as mentioned earlier [26]. The morphology of the gamma-ray emission from SNRs often exhibits a shell-like structure and is a consequence of either bremsstrahlung and inverse Compton scattering (leptonic models) or neutral pion decay (hadronic models). In the last decade, the precise measurement of gamma-ray spectra allowed to distinguish between both scenarios for various SNRs [27]. Around 14 shell-type remnants have been detected so far in the very-high-energy regime.

Compact object binary systems These binary systems consist of a compact object (black hole or neutron star) and an orbiting high-mass (OB star) or low-mass (in general less than 1 solar mass) companion [28, 29]. If gamma rays dominate the spectra of these systems, which are usually known to be one of the strongest X-ray emitters, they are referred to as gamma-ray binaries. Very-high-energy emission has only been observed from 8 high-mass gamma-ray binaries so far, for which the nature of their compact object remains mostly unknown. Only for two systems the compact object has been firmly identified as a pulsar, and thus the VHE emission is generally explained by the interaction of the pulsar wind with the stellar wind of the companion [30, 31]. However, a microquasar scenario, in which the system is powered by accretion onto a stellar-mass black hole or a neutron star, cannot be ruled out [32].

Star clusters In theory the collective effect of colliding stellar winds can accelerate particles to hundreds of TeV allowing VHE gamma-ray production of hadronic origin ([33, 34] and references therein). To date TeV emission has been observed from four star clusters, but since essentially all Galactic gamma-ray sources involve high-mass star evolution, an association with a single object within the cluster, such as a PWN, is difficult to exclude. Only recently [35] for the H.E.S.S. collaboration announced the first unambiguous detection of VHE gamma-ray emission from colliding stellar winds in the binary system Eta Carinae, which hosts a luminous blue variable ($\sim$100 $M_\odot$) and an O- or B-type star ($\sim$30 $M_\odot$).

1.2.2 Extragalactic Accelerators

Active galactic nuclei An active galactic nuclei (AGN) is a compact central region of a galaxy that harbors an actively accreting supermassive black hole ($\gtrsim 10^6 M_\odot$) outshining most of its host galaxy. Powered by accretion, these astrophysical objects eject relativistic outflows (jets) and are the most luminous nonexplosive sources in the Universe emitting over the whole electromagnetic spectrum [36]. At present, AGNs make up about $\sim$35% of the $\sim$200 sources detected in the VHE regime. The vast majority of VHE AGNs are blazars, that is their jets are closely aligned with the line of sight to Earth. Blazars typically exhibit a "double humped" Spectral Energy Distribution (SED) of continuous non-thermal emission that is thought to originate from accelerated particles in the jet. While the low energy part is generally attributed to synchrotron radiation from relativistic electrons, the nature of the second component, which extends into the VHE band, is still unclear and both, leptonic as well as hadronic models are actively debated [37]. Another defining characteristic of blazars is their strong variability, especially in the X-ray and gamma-ray band.

Starburst galaxies   In starburst galaxies gas densities in the interstellar medium (ISM) are much higher compared to our Galactic ISM and the rate at which stars are formed is greatly enhanced. Numerous massive stars and supernovae inject huge amount of kinetic energy into the ISM that can not be fully compensated for by radiative cooling, and as a consequence a starburst wind fills the galactic halo with additional gas and non-thermal particles. So far only two starburst galaxies have been detected by ground-based gamma-ray detectors. Their diffuse gamma-ray emission is thought to be primarily of hadronic origin, for which the relativistic particles are provided by supernovae. For a more detailed discussion of these systems the reader is referred to Ohm [38] and references therein.

Gamma-ray bursts   With an energy output of up to $\lesssim 10^{55}$ erg, if isotropic emission is assumed, gamma-ray bursts (GRBs) are the most luminous explosions in the Universe after the Big Bang [39]. These explosions occur at cosmological distances and typically last about $\sim$0.3 s (short GRBs, <2 s) or $\sim$30 s (long GRBs, >2 s) liberating their energy primarily in the MeV-band. The prompt gamma-ray emission is usually followed by an afterglow that has been observed from radio to X-rays and that gradually decays over a timespan of hours or days. The physical processes behind the prompt emission and the afterglow are still poorly understood, but on observational grounds, some long GRBs could be firmly associated with the core-collapse supernova events of massive stars. And only recently the multi-messenger observation of a binary neutron star merger gave proof to the nature of at least some of the short GRBs [40]. The spectral shape of GRBs at the high-energy end above $\sim$10 GeV is still elusive, since ground-based gamma-ray telescopes have still not managed to uncover a signal, despite ongoing follow-up efforts of GRB alerts for the last decade. Simulations for the future CTA observatory, however, predict a detection rate of order a few GRBs per year [41].

Another intriguing source of VHE gamma rays, and a key part of the CTA research program, could be the annihilation of dark matter particles. The true nature of dark matter is still unknown, but most popular models rest upon some kind of weakly interacting massive particle (WIMP) that, following the standard cosmological model, drives the formation of gravitational wells in which baryonic matter accumulates and where galaxies are formed ([42] and references therein). If the WIMP falls in the TeV mass range, its annihilation should lead to enhanced gamma-ray fluxes in regions of high dark matter densities, such as the center of our galaxy. However, the rich field of VHE gamma-ray sources in the central region hampers the unambiguous identification of a possible dark matter signal. Hence, dwarf spheroidal galaxies of the Local Group, in which an astrophysical gamma-ray background has yet to be discovered, are also considered prime targets for dark matter searches despite their larger distances and low masses compared to the Milky Way or the Large Magellanic Cloud [13].

For a complete list of source types and an extensive introduction to very-high-energy astrophysics, the reader is referred to Longair [43], as well as Lorenz and Wagner [44] or Hinton and Hofmann [3] and references therein.

References

1. López Coto R (2017 July) Very-high-energy γ-ray observations of pulsar wind nebulae and cataclysmic variable stars with MAGIC and development of trigger systems for IACTs . Springer International Publishing. https://doi.org/10.1007/978-3-319-44751-3. ISBN 978-3-319-44751-3.
2. Simpson JA (1983) Elemental and isotopic composition of the galactic cosmic rays. Ann Rev Nucl Part Sci 33(1):323–382. https://doi.org/10.1146/annurev.ns.33.120183.001543
3. Hinton J, Hofmann W (2009) Teraelectronvolt astronomy. Ann Rev Astron Astrophys 47(1):523–565. https://doi.org/10.1146/annurev-astro-082708-101816
4. Atwood W et al (2013) Pass 8: Toward the full realization of the Fermi-LAT scientific potential. *arXiv*
5. Kerr M (2010) Likelihood methods for the detection and characterization of gamma-ray pulsars with the Fermi large area telescope. Ph.D. thesis
6. Thompson DJ (2008) Gamma ray astrophysics: the EGRET results. Rep Progr Phys 71(11):116901. https://doi.org/10.1088/0034-4885/71/11/116901
7. Ackermann M et al (2012) The Fermi large area telescope on orbit: event classification, instrument response functions, and calibration. Astrophys J Suppl Ser 203(1):4. https://doi.org/10.1088/0067-0049/203/1/4
8. Acero F et al (2015) Fermi large area telescope third source catalog. Astrophys J Suppl Ser 218(2):23. https://doi.org/10.1088/0067-0049/218/2/23
9. Atwood WB et al (2009) The large area telescope on the Fermi gamma-ray space telescope mission. Astrophys J 697(2):1071–1102. https://doi.org/10.1088/0004-637X/697/2/1071
10. de Naurois M, Mazin D (2015) Ground-based detectors in very-high-energy gamma-ray astronomy. Comptes Rendus Phys 16(6–7):610–627. https://doi.org/10.1016/j.crhy.2015.08.011
11. Aleksić J et al (2016) The major upgrade of the MAGIC telescopes, Part II: A performance study using observations of the Crab Nebula. Astropart Phys 72:76–94. https://doi.org/10.1016/j.astropartphys.2015.02.005
12. Parsons RD et al (2015) HESS II data analysis with ImPACT. In: 34th international cosmic ray conference. The Hague, Netherlands
13. The Cherenkov Telescope Array Consortium et al (2017) Science with the Cherenkov Telescope Array
14. Amenomori M et al (2015) Search for gamma rays above 100 TeV from the Crab Nebula with the Tibet Air Shower Array and the 100 m² Muon detector. Astrophys J 813(2):98. https://doi.org/10.1088/0004-637X/813/2/98
15. Bartoli B et al (2013) TeV gamma-ray survey of the northern sky using the ARGO-YBJ detector. Astrophys J 779(1):27. https://doi.org/10.1088/0004-637X/779/1/27
16. Abdo AA et al (2009b) Milagro observations of multi-TeV emission from galactic sources in the Fermi bright list. Astrophys J 700(2):L127–L131. https://doi.org/10.1088/0004-637X/700/2/L127
17. Abeysekara AU et al (2017) The 2HWC HAWC observatory gamma-ray catalog. Astrophys J 843(1):40. https://doi.org/10.3847/1538-4357/aa7556
18. Ackermann M et al (2016) 2FHL: the second catalog of hard Fermi-LAT sources. Astrophys J Suppl Ser 222(1):5. https://doi.org/10.3847/0067-0049/222/1/5

19. Di Sciascio G (2016) The LHAASO experiment: from gamma-ray astronomy to cosmic rays. Nuclear Part Phys Proc 279–281:166–173. https://doi.org/10.1016/j.nuclphysbps.2016.10.024
20. Tluczykont M et al (2014) The HiSCORE concept for gamma-ray and cosmic-ray astrophysics beyond 10TeV. Astropart Phys 56:42–53. https://doi.org/10.1016/j.astropartphys.2014.03.004
21. Knödlseder J (2016) The future of gamma-ray astronomy. Comptes Rendus Phys 17(6):663–678. https://doi.org/10.1016/j.crhy.2016.04.008
22. Bonning EW et al (2007) Accretion disk temperatures and continuum colors in QSOs. Astrophys J 659(1):211–217. https://doi.org/10.1086/510712
23. Wakely SP, Horan D (2007) TeVCat: an online catalog for very high energy gamma-ray astronomy. In: 30th international cosmic ray conference. Merida, Mexico
24. Branch D, Wheeler JC (2017) Supernova explosions. Astronomy and Astrophysics Library, Springer, Berlin Heidelberg. https://doi.org/10.1007/978-3-662-55054-0. ISBN 978-3-662-55052-6
25. Malkov MA, Drury LO (2001) Nonlinear theory of diffusive acceleration of particles by shock waves. Rep Progr Phys 64(4):429–481. https://doi.org/10.1088/0034-4885/64/4/201
26. Drury LO (2012) Origin of cosmic rays. Astropart Phys 39–40(1):52–60. https://doi.org/10.1016/j.astropartphys.2012.02.006
27. Hewitt JW, Lemoine-Goumard M (2015) Observations of supernova remnants and pulsar wind nebulae at gamma-ray energies. Comptes Rendus Phys 16(6–7):674–685. https://doi.org/10.1016/j.crhy.2015.08.015
28. Liu QZ et al (2006) Catalogue of high-mass X-ray binaries in the Galaxy (4th edition). Astron Astrophys 455(3):1165–1168. https://doi.org/10.1051/0004-6361:20064987
29. Liu QZ et al (2007) A catalogue of low-mass X-ray binaries in the Galaxy, LMC, and SMC (fourth edition). Astron Astrophys 469(2):807–810. https://doi.org/10.1051/0004-6361:20077303
30. Dubus G (2013) Gamma-ray binaries and related systems. Astron Astrophys Rev 21(1):64. https://doi.org/10.1007/s00159-013-0064-5
31. Dubus G (2015) Gamma-ray emission from binaries in context. Comptes Rendus Phys 16(6–7):661–673. https://doi.org/10.1016/j.crhy.2015.08.014
32. Massi M et al (2017) The black hole candidate LS I+61°303. Mon Not R Astron Soc 468(3):3689–3693. https://doi.org/10.1093/mnras/stx778
33. Aharonian F et al (2007) Detection of extended very-high-energy γ-ray emission towards the young stellar cluster Westerlund 2. Astron Astrophys 467(3):1075–1080. https://doi.org/10.1051/0004-6361:20066950
34. Bykov AM (2014) Nonthermal particles and photons in starburst regions and superbubbles. Astron Astrophys Rev 22(1):77. https://doi.org/10.1007/s00159-014-0077-8
35. Leser E et al (2017) First results of Eta car observations with H.E.S.S.II. In: 35th international cosmic ray conference. Busan, South Korea
36. Padovani P et al (2017) Active galactic nuclei: what's in a name. Astron Astrophys Rev 25(1):2. https://doi.org/10.1007/s00159-017-0102-9
37. Böttcher M et al (2013) Leptonic and hadronic modeling of Fermi-detected blazars. Astrophys J 768(1):54. https://doi.org/10.1088/0004-637X/768/1/54
38. Ohm S (2016) Starburst galaxies as seen by gamma-ray telescopes. Comptes Rendus Phys 17(6):585–593. https://doi.org/10.1016/j.crhy.2016.04.003
39. Kumar P, Zhang B (2015) The physics of gamma-ray bursts & relativistic jets. Phys Rep 561(Oct 2014):1–109. https://doi.org/10.1016/j.physrep.2014.09.008
40. Abbott BP et al (2017) Multi-messenger observations of a binary neutron star merger. Astrophys J 848(2):L12. https://doi.org/10.3847/2041-8213/aa91c9
41. Inoue S et al (2013) Gamma-ray burst science in the era of the Cherenkov Telescope Array. Astropart Phys 43:252–275. https://doi.org/10.1016/j.astropartphys.2013.01.004

42. Bertone G et al (2005) Particle dark matter: evidence, candidates and constraints. Phys Rep
    405(5–6):279–390. https://doi.org/10.1016/j.physrep.2004.08.031
43. Longair MS (2011) High energy astrophysics, 3rd edn. Cambridge University Press, Cam-
    bridge. ISBN 0521756189
44. Lorenz E, Wagner R (2012) Very-high energy gamma-ray astronomy. Eur Phys J H 37(3):459–
    513. https://doi.org/10.1140/epjh/e2012-30016-x

Chapter 2
Pulsars and Pulsar Wind Nebulae

Unusual signals from pulsating radio sources have been recorded at the Mullard Radio Astronomy Observatory. The radiation seems to come from local objects within the galaxy, and may be associated with oscillations of white dwarf or neutron stars.

Hewish et al., 1968

When heavy stars have burned all their nuclear fuel, neutron degeneracy pressure is the last force able to halt their collapse into a black hole. The sudden stop of the free-fall collapse leads to a rebound of the infalling matter triggering an outward shock that blows up the star envelope and powers a *Type II* supernova. If the mass of the progenitor star does not exceed ~20 solar masses [1], the compact remnant core evolves into a *neutron star*. Otherwise the amount of matter falling back on to the core crosses the maximum neutron star mass and ultimately collapses to form a black hole.

Neutron stars are about 24 km in diameter and have masses of approximately 1.5 solar masses [2]. Conserving both the angular momentum and the magnetic flux of the progenitor star, these compact objects spin at a frequency of up to ~700 Hz and exhibit magnetic field strengths of 10^8–10^{15} G. The acceleration of charged particles in these extreme conditions gives rise to powerful beams of light that sweep the universe like a light house (Fig. 2.1). If such a beam hits the earth, we are able to detect the neutron star as a *pulsar* (portmanteau of *pulsating stars*) that emits a periodic signal at a precision comparable to the accuracy of atomic clocks. The total amount of energy released by a typical pulsar over its lifetime is about ~10^{50} erg, but only $\lesssim 10\%$ of this energy is emitted as beamed electromagnetic radiation [3]. Most of the energy is released in form of a pulsar wind consisting of relativistic particles and magnetic fields. The pulsar wind can feed a *Pulsar Wind Nebula* (PWN), which is a cloud of magnetized electron-positron plasma extending up to ~40 pc away from the neutron star [4]. Pulsar wind nebulae are typically observed by their synchrotron radiation

© Springer Nature Switzerland AG 2019
D. Carreto Fidalgo, *Revealing the Most Energetic Light from Pulsars and Their Nebulae*, Springer Theses, https://doi.org/10.1007/978-3-030-24194-0_2

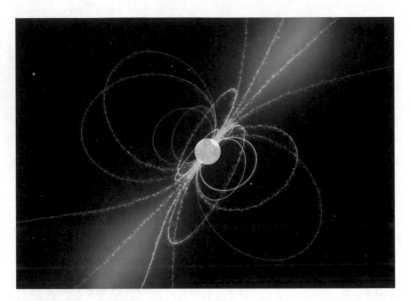

Fig. 2.1 Artistic view of a pulsar. The red hot sphere represents the neutron star, the blue lines the strong magnetic field and the magenta haze depicts the beamed emission that sweeps the universe like a light house. Figure taken from https://www.nasa.gov/mission_pages/GLAST/news/gr_pulsar. html, last accessed 11/08/2017 (Color figure online)

in the radio and X-ray bands, but are also able to produce TeV photons via inverse Compton (IC) scattering, predominantly on the cosmic microwave background (CMB) and infra-red photon seed fields.

In this chapter we will provide an introduction to pulsar and pulsar wind nebula physics, focusing on the observational properties and emission mechanisms of gamma rays in the GeV to TeV range. The first three sections deal with the observational variety of neutron stars, the magnetosphere of a pulsar and some spin-down properties, respectively, while Sect. 2.4 will introduce the class of gamma-ray pulsars and discuss their current emission models. From Sect. 2.5 on we will talk about PWNe providing a short overview of their general evolution and summarize observational results on TeV pulsar wind nebulae in the last section.

2.1 Neutron Star Zoo

Neutron stars are one of the few objects in astronomy that were successfully predicted before their actual discovery. Soon after the identification of the neutron in 1932 [5], Baade and Zwicky [6] proposed that the supernova process could mark the transition of an ordinary star into a neutron star. Subsequently further publications explored

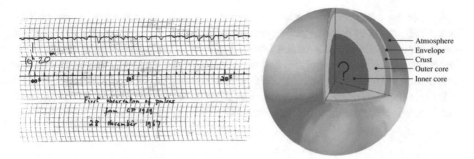

Fig. 2.2 Left: Historical chart record of the first pulsar, CP1919, discovered in 1967. The periodic pulses are observed as dips in the upper trace spaced by ~1.3 s. Picture taken from http://www.cv. nrao.edu/~sransom/web/Ch6.html, last accessed 07/08/2017. **Right**: The interior of a neutron star can be divided into five major regions (see text). Image taken from [8]

the idea of exotic stars with degenerated cores and important theoretical groundwork were laid by R. C. Tolman, J. R. Oppenheimer and G. M. Volkov in the late 40s. Still today their equations constitute the base of our knowledge about composition and structure of neutron stars, such as the mass-radius relation and the relation between pressure and mass/energy density, also called the equation of state [2]. In 1967 the discovery of the first pulsar by Jocelyn Bell [7], finally delivered the smoking gun to the existence of these exotic stars (see the left image of Fig. 2.2). The following findings of short-period pulsars ($P \simeq 50$ ms) and the gradual slow-down of the periodic signal from the Crab pulsar made the association with a rapidly spinning neutron star certain.

2.1.1 The Interior of a Neutron Star

The interior composition and structure of a neutron star is still highly uncertain and an active research topic. Here we will provide a brief description of the five major regions sketched in Fig. 2.2 following Lattimer and Prakash [2] and Page and Reddy [8]. At the neutron star's surface a very thin atmosphere is expected mainly composed of hydrogen. From the observational point of view the atmosphere plays a key role in determining the temperature of a neutron star since it is able to cause significant deviations to an assumed blackbody spectrum. Below the atmosphere not yet fully degenerated matter forms an *envelope* of a few tens of meter. The envelope acts as a thermal insulator between the hot interior of the star and the surface. The outer 500–1000 m of the interior contains nuclei in a lattice immersed in a quantum liquid of neutrons and is referred to as crust. The dominant nuclei in the crust vary with density and range from ^{56}Fe to extremely neutron-rich nuclei of ~200 nucleons. Changes in the crust are believed to be the main reason for sudden glitches in the rotation of a neutron star as is often observed in young pulsars. As density grows towards the center, the crust also marks a phase transition from the inhomogeneous regime of the outer crust to the homogeneous core (also termed *nuclear pasta*). The outer core starts at a density of approximately 0.6 of the nuclear density where a neutron superfluid coexists with a proton fluid and a soup of electrons and muons. In the inner core exotic particles and novel phases could emerge that dominate the interaction within the particles and affect the behavior of the core, which makes up ~99% of the total mass of the neutron star. For a more detailed review of the neutron star's interior composition and structure, the reader is referred to [8].

2.1.2 Neutron Star Classes

One of the common observational characteristics of all neutron stars is typically the detection of a pulsating signal from the star due to its rotation. Otherwise they exhibit surprisingly diverse observational and physical properties. The neutron star zoo can be classified according to the primary power source for their emission and spin evolution. In the following we provide an overview of the different populations following [9].

The biggest class of neutron stars are the rotation-powered pulsars (RPPs), which spin down as a result of torques from magnetic dipole radiation and particle emission. Their emission appears as broad-band pulsation from radio to gamma-rays wavelengths. There are two main populations of RPPs: (i) canonical pulsars with rotational periods from tens of milliseconds to tens of seconds, which have a characteristic age of < 100 Myr; (ii) Millisecond Pulsars (MSPs) with periods of a few milliseconds and a characteristic age of $\gtrsim$100 Myr (see lower left population in Fig. 2.3). Millisecond pulsars make up about 10% of the RRPs and about 80% of them are in binary systems [9]. It is widely accepted that MSPs have originally been members of the canonical RPP population and at some point in time were spun up (or "recycled") by accretion from a binary companion. Solid proofs of the evolutionary link between accretion and rotation-powered MSPs were found in the last decade by, for example, Archibald et al. [10] and Papitto et al. [11]. The most studied object of the canonical type is arguably the Crab pulsar, which will also be the main topic of this thesis.

The class of neutron stars called *magnetars*, primarily draw the energy from their magnetic fields of unprecedented strength of $10^{14} - 10^{16}$ G. Magnetars show steady X-ray pulsations as well as soft gamma-ray bursts and can be subclassified in Anomalous X-Ray Pulsars (AXPs) and Soft Gamma-Ray Repeaters (SGRs), respectively. The inferred steady X-ray luminosities exceed the spin-down luminosities by a factor of about 100 requiring a source of power other than their rotational energy.

Central Compact Objects (CCOs) are X-ray sources with no counterparts in other wavebands and are detected near the center of young supernova remnants. For some of the CCOs pulsation in the hundreds of ms range has been detected with low $\dot{P}$ resulting in low magnetic fields in the range of $10^{10} - 10^{11}$ G. They exhibit exclusively thermal spectra, some with multiple blackbody components, indicating that the emission is from cooling of the neutron star.

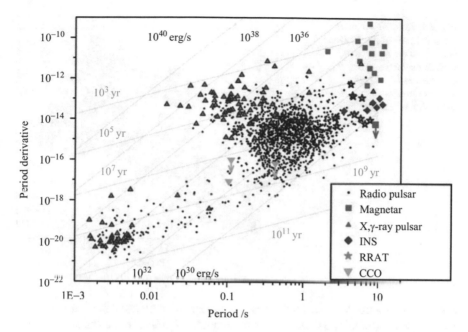

Fig. 2.3 Neutron stars (NS) for which pulsation is detected, plotted in their period versus period time derivative space. The different NS/pulsar classes are given in the legend. For a short description of each class, see the text. Lines of constant characteristic age (light blue) and dipole spin-down luminosity (black) are superposed (see Sect. 2.3 for a definition of both quantities). In the $P - \dot{P}$ plane one is able to clearly distinguish the millisecond pulsar population (bottom left) from the canonical population. Plot taken from [9] (Color figure online)

Another class that appears to be thermally cooling with no emission outside the soft X-ray band, are the Isolated Neutron Stars (INS). In contrast to the CCOs they lack any associated supernova remnant or nebula. The measured X-ray pulsation has periods of 3–11 s, much longer than those of CCOs, and their surface magnetic fields of around 10^{13} G are larger.

The population of accreting neutron stars exploit the gravitational energy of infalling matter from a companion star and their spin evolution is dominated by the accretion torques. Accretion can either occur from the stellar winds of the companion or from an accretion disk if the companion overflows its Roche Lobe. Depending on the mass of the companion star and the type of accretion, these neutron stars display a variety of physical and observational behaviors and are grouped in subclasses accordingly. Binary systems in which the companion is a low-mass main sequence star, white dwarf or red giant that fills its Roche Lobe, are called Low-Mass X-Ray Binaries (LMXBs), since basically all of their luminosity (~99%) is emitted in X-rays. Neutron stars in LMXBs are thought to be spun up by torques from the accretion disk until they reach an equilibrium period in the range of milliseconds. X-ray pulsation at millisecond periods are in fact observed in a number of LMXBs (for example [11], and therefore LMXBs are thought to be the intermediate stage in the recycling

Fig. 2.4 Rough location of the subclasses of the neutron star zoo in the period and surface magnetic field space. See the text for a short description of each one of them. Plot taken from [9]

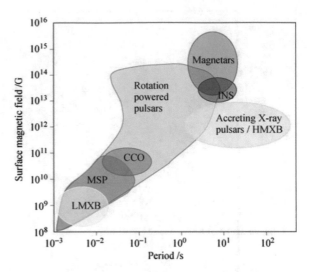

scenario to produce MSPs. In High-Mass X-Ray Binaries (HMXBs) the neutron star or black hole is orbited by a O or B star with a mass of $\gtrsim 5M_\odot$. In HMXBs the accretion happens either from the stellar wind and outbursts of the companion or from a Be star disk. Another subclass of accreting neutron stars are the *microquasars* that show rapid variability of their X-ray emission, radio jets and strong broadband emission with broad emission lines.

Figure 2.4 gives a summary of the different classes by plotting rough contours in the period and surface magnetic field space.

2.2   Pulsar Magnetosphere

While the pulsed nature of the signal from neutron stars is a direct consequence of the star's rotation, there are still doubts about how the signal is produced in the first place. A key role in the production of beamed photons is surely played by the pulsar's magnetosphere. The magnetosphere is defined as the spatial region surrounding the neutron star in which the electromagnetic forces dominate over all other forces, such as gravitational or pressure-gradient ones. It is generally believed that the magnetosphere is filled with magnetized plasma due to the extreme magnetic fields of the neutron star, as we will demonstrate further down in this section.

2.2.1   Force-Free Configuration

To model a general and consistent configuration of the pulsar's magnetosphere is an exceedingly complex task involving many macro- and microphysical variables, such as the large-scale magnetic field configuration or the generation of plasma and electromagnetic radiation.

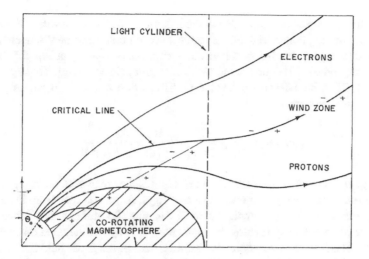

Fig. 2.5 Sketch of the pioneering model for the pulsar electrodynamics by [12]. The rotation and dipole axis of the pulsar are aligned in the y-direction (see bottom left corner). In this sketch the feet of the *critical* field lines, are at the same electric potential as the interstellar medium, and thus electrons escape along higher-altitude lines, while protons do at lower latitudes. Figure taken from [12]

A pioneering model for the pulsar electrodynamics was proposed by Goldreich and Julian [12] who assumed the basic case of a rotating magnetic dipole aligned with the rotation axis of the star (see Fig. 2.5). In their picture electric fields arise via the unipolar induction mechanism, the same principle Michael Faraday used to develop the first homopolar generator (also called Faraday disc), which converts kinetic energy into electric voltage. Goldreich and Julian [12] were also the first to point out that these strong electric fields parallel to the magnetic field lines, will exceed gravitational forces by many orders of magnitude and inevitably rip off charged particles from the neutron star surface. In a vacuum, those particles are further accelerated and provoke electromagnetic cascades, which renders a vacuum around the pulsar unstable. As a consequence [12] predicted a dense magnetized plasma around the pulsar acting as a perfect conductor under the condition that the neutron star provides a free supply of charged particles. Perfect conduction implies that the system can separate charges q to compensate for the magnetic force, that is $(\mathbf{E} + \mathbf{v} \times \mathbf{B})\, q = 0$, and hence $\mathbf{E} \cdot \mathbf{B} = 0$ (often referred to as the ideal magnetohydrodynamic, MHD, condition), where $\mathbf{E}$ and $\mathbf{B}$ denote the electric and magnetic field, respectively, and $\mathbf{v}$ the velocity vector of the particle. This configuration is known as the force-free electrodynamic (FFE) or force-free configuration of the magnetosphere and is still often used to some extent in modern magnetospheric and pulsar emission models.

The magnetized plasma is dragged around by the strong magnetic field lines and corotates with the neutron star up to the so-called light cylinder, an imaginary cylinder around the star where the tangential velocity reaches the speed of light. As indicated in Fig. 2.5 the corotating magnetosphere (also called closed magnetosphere) is confined within the last closed field line, that is the last magnetic field line which lies entirely within the light cylinder. The feet of the last closed field lines on the neutron star's surface encompass the region called polar cap, denoted as θ_0 in Fig. 2.5. The FFE condition allowed [12] to derive the *Goldreich-Julian* space charge density ρ_{GJ} (see Eq. 8 in [12]), which describes the charge distribution within the corotating portion of the magnetosphere:

$$\rho_{GJ}(r, \theta) = -\frac{\Omega \cdot \mathbf{B}}{2\pi c \left(1 - (\Omega r \sin \theta / c)^2\right)} \quad .$$

(2.1)

ρ_{GJ} is given in polar coordinates where θ denotes the angle with the rotation vector Ω of the pulsar and c represents the speed of light. As can be seen from Eq. 2.1, ρ_{GJ} equals zero when $\mathbf{B}$ is perpendicular to Ω forming the null charge surface, where the space charge density reverses its sign (indicated by the diagonal dashed line in Fig. 2.5). The FFE magnetosphere also implies that the magnetic field lines are at the same time equipotential contours allowing the charged particles to move away from the neutron star. Current outflows from the pulsar surface into the interstellar space are possible where the magnetosphere exhibits open magnetic field lines, dubbed as the open magnetosphere in contrast to the closed one. The sign of the current outflows depends on the electric potential of the interstellar medium and a critical field line is defined where the potentials match each other (see Fig. 2.5). Together with the pointing flux emitted by the magnetic dipole radiation, these current outflows form the so-called *pulsar wind* that energizes the surrounding environment and can lead to the formation of a pulsar wind nebula. The pulsar wind also carries energy and angular momentum away from the pulsar, providing a braking torque for the rotation of the neutron star, as we will discuss in the next section.

In the simple case of an aligned rotator with a FFE magnetosphere, a self-consistent description of the fields and currents is given by the so-called pulsar equation [14, 15]. The equation is applicable to any magnetic multipole case, but has only an exact analytical solution in the case of a split monopole field configuration. In this configuration the magnetic field lines resemble the ones of a magnetic monopole, pointing towards the pulsar on one hemisphere and away from the pulsar on the other one [14]. A numerical solution to the pulsar equation assuming a magnetic dipole was first found by Contopoulos et al. [16]. Their results were confirmed by Timokhin [13] who also showed the formation of a current sheet along the spin equator closing the magnetospheric current that flows out of the polar regions. The return current reaches the neutron star along the separatrices forming a Y-configuration together with the current sheet as shown in Fig. 2.6. One of the main results of these numerical solutions was that the relativistic outflow along the open magnetic field lines is not accelerated to energies required to generate gamma rays. [17] went a step further and presented the three-dimensional structure of the magnetosphere in case of an oblique rotator, by numerically solving the time-dependent Maxwell's equations together with the force-free condition (see Fig. 2.7). While FFE magnetoshperes are useful to obtain global configurations of the magnetic field, charges and currents, they can give us only limited insight about particle acceleration and emission, since

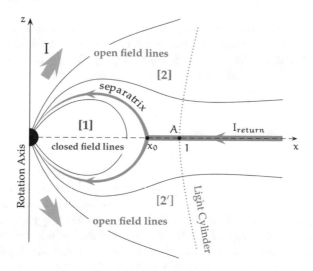

Fig. 2.6 Configurations of the force-free magnetosphere in the case of an aligned rotator and the macroscopic current closure. The current flows out in the open field line regions denoted as [2] and [2′], closes somewhere beyond the light cylinder and returns along a current sheet and the separatrix. In Sect. 2.5 we will see that, for an oblique rotator, the current sheet undulates around the equatorial plane tracing stripes. The Y-configuration of the return current is apparent with the Y null point denoted as x_0. Figure taken from [13]

Fig. 2.7 3D magnetic field line configuration of an oblique rotator (60°) with the force-free condition. The color represents the toroidal field component. A sample flux tube is traced in white. Figure taken from [17] (Color figure online)

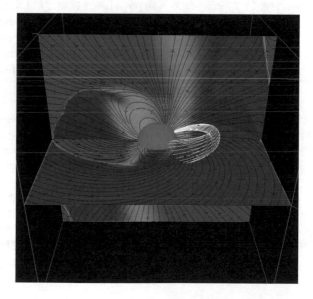

they demand by definition a vanishing electric field along the magnetic field lines. Currents perpendicular to the field lines are generally neglected because the strong magnetic fields lead to very short synchrotron cooling times (much shorter than the pulsar period) and most of the charges must be in the lowest Landau state [18].

2.2.2   Dissipative Magnetospheres

In the past few years efforts have been made to develop dissipative pulsar magneto-sphere models, where the ideal MHD condition is dropped [19, 20]. In these models a formulation of Ohm's law is chosen to relate the macroscopic current to the electric field $E_\parallel$ along the magnetic field lines with a global conductivity parameter σ. These resistive solutions smoothly bridge the gap between the vacuum and the force-free magnetosphere. In the resistive solutions, high absolute values for $E_\parallel$ arise primarily above the polar caps, near the separatrices and near the equatorial current sheets out-side of the light cylinder (see Fig. 2.8). Kalapotharakos et al. [21] and Brambilla et al. [22] computed pulsar light curves for a range of such dissipative magnetospheres, by simulating realistic trajectories of particles starting from the stellar surface around the polar cap and taking into account their emission due to curvature radiation. When comparing the obtained radio lags[1] and the gamma-ray peak separations to the ones observed by *Fermi*-LAT, Kalapotharakos et al. [21] found that their so-called FIDO (force-free inside, dissipative outside) model is the best in reproducing the exper-imental data. In this hybrid model, they assumed a FFE solution inside the light cylinder ($\sigma \to \infty$) and a high, but finite, σ outside the light cylinder ($\sigma = 30\Omega$). The success of these dissipative MHD magnetosphere models, however, comes with the price of lacking a fundamental self-consistency. In the MHD approach the source of the charges is not specified explicitly and it is assumed that charges are supplied freely and are sufficient to support whatever macroscopic current. Also the amount of dissipation, the conductivity parameter σ, is defined ad hoc based on observational requirements, instead of relying on microphysical processes.

In this respect, N-body or particle-in-cell (PIC) simulations are a superior approach to construct a fully self-consistent pulsar magnetosphere. The drawback of such simulations is computational, since particle codes in general are considerably more demanding when compared to MHD simulations. Recently Philippov et al. [23] studied the formation of a pulsar's magnetosphere by means of relativistic PIC simulations, in which the plasma is not only provided by the surface of the neutron star but also from pair production[2] by high-energy photons. As shown in Fig. 2.9 the results resemble closely the ideal force-free magnetosphere with the formation of the Y-point and the

---

[1]The radio lag refers to the phase lag between the first gamma-ray peak and the main radio peak (a definition of *pulse phase* can be found in Appendix B). For the computation of the radio lag, Kalapotharakos et al. [21] assumed that the radio emission originates from near the magnetic pole on the pulsar surface.

[2]The pair production mechanism in their study is highly idealized and is limited to a pair multiplicity of 10. For details see [23].

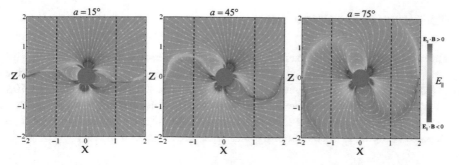

Fig. 2.8 Magnetic field line configuration of various oblique rotators in the poloidal plane (dipole axis μ, rotation axis Ω) obtained in a dissipative pulsar magnetosphere. The stream lines represent the magnetic field. The electric field $E_\parallel$ along the magnetic field lines is indicated by the color scale. The simulations were conducted assuming a high conductivity (near-FFE) and three different inclinations α as indicated in the title of each panel. The rotation vector Ω lies along the Z axis. Z and X are given in units of the light cylinder R_{lc} (vertical dashed line). $E_\parallel$ arises primarily above the polar caps, along the separatrix and in the current sheet. Figure taken from [21] (Color figure online)

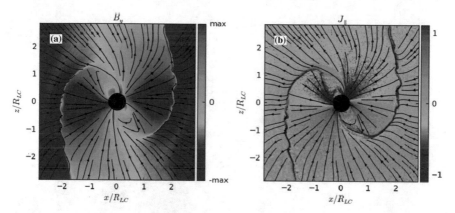

Fig. 2.9 Magnetic field configuration of an oblique rotator (60°) obtained from relativistic PIC simulations. The stream lines represent the magnetic field lines in the poloidal plane, the rotation vector lies along the Z axis. In the left panel (marked as c), the color scale indicates the toroidal (out-of-plane) component of the magnetic field. In the right panel (marked as d), the color scale denotes the normalized current along the magnetic field lines. Figure taken from [23] (Color figure online)

current sheet. They find that in solutions with low obliquities ($\leq 40°$), pair formation only happens in the current sheet and in the return current layer. For higher obliquities the pair production site includes the open field line region above the poles and the size of the pair production region increases with inclination. With a steady increase in computing power and resources, in the near future these simulations should feature a more realistic pair injection from full particle cascades and provide a next step in fully understanding the geometry and characteristics of a pulsar's magnetosphere and its accelerating regions.

2.3   Spin-Down Properties

As shortly mentioned in Sect. 2.1, pulsars show a steady slow-down of their signal's frequency. By measuring the pulse period and the slow-down, one can derive important properties of the pulsar applying some basic assumptions (see Fig. 2.10). Let us first have a look at the energetics. Approximating the pulsar with radius R as a magnetic dipole, which rotates uniformly at an angular frequency Ω, its time dependent dipole moment can be expressed as (see Eq. 10.5.3 in [26])

$$\mathbf{m}\,(t) = \frac{1}{2}B_0 R^3 \left(\sin\alpha \,\cos\Omega t \,\mathbf{e_x} + \sin\alpha \,\sin\Omega t \,\mathbf{e_y} + \cos\alpha \,\mathbf{e_z} \right), \qquad (2.2)$$

where B_0 is the magnetic field strength at the poles and α the angle between the rotational axis and magnetic dipole. The loss of energy of a varying magnetic momentum is described by the Larmor formula (see [27]) $\dot{E} = \left(2/3c^3\right)|\ddot{\mathbf{m}}|^2$, where c is the speed of light. Using Eq. 2.2 we find

$$\dot{E} = \frac{B_0^2 R^6 \Omega^4}{6c^3} \sin^2\alpha. \qquad (2.3)$$

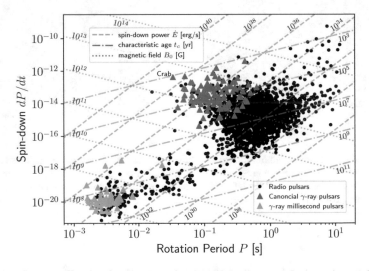

Fig. 2.10 All pulsars listed in the ATNF catalog ([24], http://www.atnf.csiro.au/research/pulsar/psrcat/, last accessed 04/09/2017), plotted in the $P - \dot{P}$ space. Gamma-ray pulsars are depicted as colored triangles (canonical as blue, millisecond as red). The spin-down properties discussed in this section, are indicated by colored lines in the background (see legend). As discussed in Sect. 2.4, gamma-ray pulsars seem to cluster around young and energetic pulsars (excluding MSPs). The most recent discovered gamma-ray pulsars are still missing in the ATNF catalog and are not plotted here. An updated list of gamma-ray pulsars can be found here: https://confluence.slac.stanford.edu/display/GLAMCOG/Public+List+of+LAT-Detected+Gamma-Ray+Pulsars (Color figure online)

Due to the low frequency Ω at which the energy is radiated away, the corresponding electromagnetic fields are able to accelerate particles efficiently as shown by Gunn and Ostriker [28]. When taking into account the magnetosphere in a force-free configuration, Spitkovsky [17] found that this energy loss can be well approximated by

$$\dot{E}_{ff} = \frac{B_0^2 R^6 \Omega^4}{4c^3} \left(1 + \sin^2 \alpha\right). \tag{2.4}$$

In the case of resistive solutions for pulsar magnetospheres, Li et al. [19] show that these solutions form a monotonic transition from force free to vacuum with decreasing conductivity (see Fig. 2.11). If the energy is provided by the rotational kinetic energy, the energy balance can be written as

$$I\Omega\dot{\Omega} = \frac{B_0^2 R^6 \Omega^4}{6c^3} f_\alpha, \tag{2.5}$$

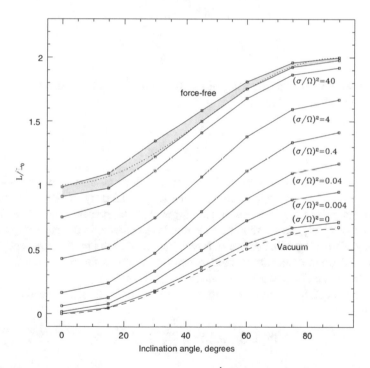

Fig. 2.11 The normalized spin-down luminosity L (or $\dot{E}$) in function of the inclination angle and for force-free, a sequence of resistive, and vacuum dipoles. The grey band of the force-free solution indicates the uncertainty in the measurement due to boundary and numerical effects. The dotted line inside the band denotes the analytical relation $\dot{E} \propto (1 + \sin^2 \alpha)$. The dashed *Vacuum* line denotes the analytical vacuum solution by [25]. Figure taken from [19]

where I is the moment of inertia and f_α is a factor, which contains the dependency on the inclination angle and reads $\sin^2 \alpha$ in the case of vacuum and $1.5 \left(1 + \sin^2 \alpha\right)$ in the case of the force-free solution. Equation 2.5 neglects other possible mechanism for the spin-down, such as the particle wind or gravitational wave emission (see and references there in [29, 30]). The derivative of the diminishing rotational kinetic energy is often referred to as spin-down luminosity or spin-down power of a pulsar and can be written as

$$\dot{E}_{sd} = I\Omega\dot{\Omega} = 4\pi^2 I \frac{\dot{P}}{P^3}, \tag{2.6}$$

where $P = 2\pi/\Omega$ is the period of the pulsar and $\dot{P}$ its time derivative, respectively. From the energy balance we can also estimate the age of a pulsar by integrating Eq. 2.5 over time,

$$\int_0^T dt = \frac{6Ic^3}{B_0^2 R^6 f_\alpha} \int_0^T \Omega^{-3} \dot{\Omega} dt \tag{2.7a}$$

$$T = \tau_c \left(1 - \frac{\Omega^2}{\Omega_0^2}\right), \tag{2.7b}$$

where Ω_0 is the angular frequency at $t = 0$ and the so-called characteristic age (or spin-down age) τ_c is defined as

$$\tau_c \equiv \frac{6Ic^3}{B_0^2 R^6 f_\alpha 2\Omega^2} = \frac{\Omega}{2\dot{\Omega}} = \frac{P}{2\dot{P}}. \tag{2.8}$$

τ_c is only a good estimate of the real pulsar age if three assumptions are fulfilled: (i) the braking torque for the spin-down is mainly provided by the Poynting flux; (ii) the initial angular frequency was much bigger than the present one ($\Omega_0 \gg \Omega$); (iii) the factor $I/B_0^2 R^6 f_\alpha$ is constant in time, that is the change in the moment of inertia, the decay of the magnetic field and a possible change in the inclination is negligible. Equation 2.8 often overestimates the true pulsar age by a factor up to a few, which may indicate that Ω_0 is usually not much bigger than Ω.

Inverting Eq. 2.5 also allows us to estimate the strength of the magnetic field at the poles

$$B_0 = 6.4 \times 10^{19} \sqrt{\frac{I_{45}}{R_6^6 f_\alpha}} \sqrt{P\dot{P}} \; G, \tag{2.9}$$

where $I_{45} = I/10^{45}$ g cm^2 and $R_6 = R/10^6$ cm. In literature the values $I_{45} = R_6 = f_\alpha = 1$ are commonly used, assuming an orthogonal rotator in a vacuum [24]. In addition one often states the magnetic field strength at the equator, rather than at the poles, which differs by a factor of $1/2$.

A way to describe the rotational evolution of a pulsar is to define a torque equation, $\dot{\Omega} = -k\Omega^n$, where k is a constant and n is the so-called braking index of a pulsar.

The braking index is a measurable quantity and can be written as

$$n = \frac{\Omega\ddot{\Omega}}{\dot{\Omega}^2} = 2 - \frac{P\ddot{P}}{\dot{P}^2} \quad . \tag{2.10}$$

If the slow-down is solely due to magnetic dipole radiation, as suggested by Eq. 2.5, the braking index n equals 3. A braking torque provided exclusively by gravitational wave emission yields an index of 5 [30]. Most pulsars, however, show braking indexes smaller than 3, which can be explained by time dependences of the magnetic field and inclination angle, or particle outflows in the magnetosphere among other mechanisms (and references there in [29]).

2.4   Gamma-Ray Pulsars and Their Emission Mechanisms

Since their discovery in 1967 over 2500 pulsars have been detected in the radio band so far [24]. Pulsations in the optical, infrared and ultraviolet, have been found only for a few tens of isolated neutron stars owing to their intrinsic faintness and strong optical extinction at low Galactic latitudes [31, 32]. In the X-ray band around 100 RPPs are known to emit periodic pulsation [33]. The X-ray emission from rotation-driven pulsars is mostly a mixture of non-thermal and thermal processes that also can exhibit a continuous component. However, most of the neutron stars observed in X-rays (several hundreds) are found in X-ray binary systems gaining their power from accretion rather than from the loss of rotational energy. In these systems almost all of the radiation ($\sim$99%) is emitted in the $\sim (5-30)$ keV range [9].

2.4.1   Observational Revolution

In the case of gamma rays, the last decade bore witness to a observational revolution due to the launch of the *Fermi* Gamma-Ray Space Telescope in 2008, which carried on board the Large Area Telescope (LAT, [34]). While prior to the launch only six hard gamma-ray pulsars (above 100 MeV) were known [35], the *Fermi* LAT managed to increase this number by more than tenfold, to 211 detections at present.[3] Pulsars represent thereby the largest Galactic source class in the gamma-ray band (see Fig. 2.12). When plotting the population of gamma-ray pulsars in the $P - \dot{P}$ diagram, as done in Fig. 2.10, a striking feature is the clustering of detections around young and energetic pulsars (excluding millisecond pulsars). This confirms the EGRET results (Energetic Gamma Ray Experiment Telescope, the precursor to the *Fermi* LAT), as well as theoretical predictions, that indicated that young ($\tau \lesssim 100$ kyr) pulsars with

---

[3]https://confluence.slac.stanford.edu/display/GLAMCOG/Public+List+of+LAT-
Detected+Gamma-Ray+Pulsars, last accessed 06/04/2018.

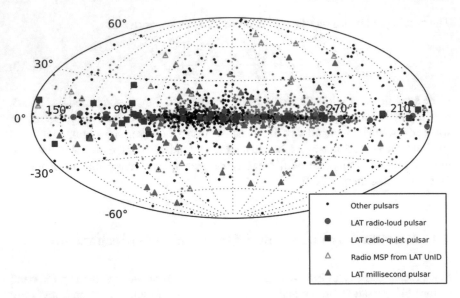

Fig. 2.12 The sky-map of known pulsars in Galactic coordinates. Grey and black points mark radio pulsars. Canonical gamma-ray pulsars are subdivided between radio-loud (green dots) and radio-quiet (blue squares). Orange open triangles indicate radio MSPs discovered at the position of previously unidentified gamma-ray sources, but for which gamma-ray pulsation was not detected yet when the 2PC was published [3]. Red triangles denote gamma-ray MSPs. The population of MSPs is less confined to the Galactic plane, which is expected due to their age and proper velocity (see Sect. 2.5). Figure taken from [3] (Color figure online)

spin-down powers $\dot{E} > 10^{34}$ erg s^{-1} are the most likely candidates to emit hard gamma-rays [36]. Another result from the population study conducted by [3] using the second *Fermi*-LAT Pulsar Catalog (2PC), settled the long-standing debate on the origin of the gamma rays in the inner or outer parts of the pulsar magnetosphere. The strong magnetic field above the neutron star's polar caps prevents the escape of photons with an energy above ε_{max} due to magnetic pair creation and photon splitting [37, 38]. Close to the stellar surface such processes thus yield steep, super-exponential[4] cut-offs in the gamma-ray spectra above a few GeV, which have not been observed so far. In fact, the vast majority of gamma-ray pulsars seen by *Fermi*-LAT exhibit simple exponential cut-offs, with a few showing an even more gradual sub-exponential cut-off. In addition, the substantial fraction of radio-quiet gamma-ray pulsars, as well as its anti-correlation with the spin-down power, strongly favors gamma-ray beams that arise in the outer magnetosphere, as shown by [39]. In this picture radio-quiet gamma-ray pulsars are a result of larger solid angles of the gamma-ray beam compared to the radio beam (see Fig. 2.13). Also the preponderance of light

---

[4]The exponential cut-off in a spectrum is described by the functional relation $dN/dE \propto e^{(-E/E_c)^b}$, where E_c defines the cut-off energy. For a simple or pure cut-off, b equals 1, whereas $b > 1$ and $b < 1$ are denoted as super-exponential and sub-exponential cut-offs, respectively.

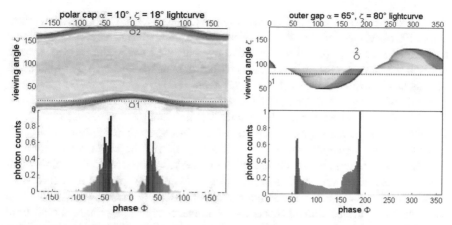

Fig. 2.13 Sky maps of pulsar beams and sample pulsar light curves for two pulsar emission models. In the sky maps the intensity of the beams (color) is plotted in function of the viewing angle (the angle between the rotation axis of the pulsar the line-of-sight from Earth) and the pulse phase (see Appendix B for a definition of pulse phase). **Left:** In the polar cap (PC) model the radiation beams originate from right above the stellar surface. The sky map is drawn for a PC model with an inclination of 10°. The PC model produces two narrow beams that are completely missed for most of the viewing angles. The example light curve is plotted for a viewing angle of 18°. **Right:** In the outer gap (OG) model the beams are produced in the outer part of the magnetosphere. The OG model produces wider beams that are more likely to hit the earth. The sharp caustic peaks of the OG model, as often observed in gamma-ray pulsars, are illustrated in an example light curve for a viewing angle of 80°. Figures taken from [41]

curves with two significantly separated caustic peaks and the observed radio lags[5] is another indication for wide gamma-ray emission beams originating in the outer part above the null charge surface [40].

In general the exponential GeV cut-off in gamma-ray pulsars can be explained by curvature radiation (CR) at the radiation-reaction limit, where the acceleration of the particle is balanced by CR losses (see for example [42]). In this regime the maximum photon energy is determined by the accelerating electric field along the magnetic field lines and their curvature radius. Sub-exponential cut-offs, as seen in some of the phase averaged pulsar spectra, can conceptually be explained by the superposition of several pure exponential shapes with a range of cut-off energies. Such a range of cut-off energies is indeed observed in phase resolved analysis of bright gamma-ray pulsars such as Vela or Geminga [43, 44]. In theory a range of cut-off energies can also be achieved at the same phase, since the caustic peaks blend the emission from different spatial regions of the magnetosphere [45] or since non-stationary emission models could blend variable curvature radiation in time [46].

As a real surprise came the large fraction of MSPs among the gamma-ray pulsars seen by *Fermi*-LAT. While the precursor to the *Fermi* LAT, EGRET, was not able to

---

[5]The radio lag refers to the phase lag between the first gamma-ray peak and the main radio peak.

detect a single MSP in the hard gamma-ray band,[6] there are now 93 firm detections
above 100 MeV, most of them found in binary systems. Hence, MSPs constitute almost
half of the present population of gamma-ray pulsars. Due to the weak magnetic
fields at the neutron star surface (at least two orders of magnitude lower than those
of canonical pulsars, see Fig. 2.10), MSPs were thought to have certain issues in
producing rich electron-positron pair cascades and screening the accelerating electric
field [48, 49]. However, the similar spectra and light curves compared to canonical
pulsars suggest that the gamma-ray emission from MSPs also originates from the
outer part of their magnetospheres. And indeed, most of the MSP gamma-ray light
curves are well fit with outer magnetosphere models as shown by [50], implying that
these old pulsars are still capable of providing sufficient pair plasma for a force-free
condition throughout most of the magnetosphere. Despite their low magnetic field
strength at the surface, some MSPs exhibit comparatively strong fields at their light
cylinder, due to their high spin frequency. Interestingly, this makes MSPs promising
candidates for VHE synchrotron self-Compton (SSC) emission as pointed out by [51].
The relatively small corotating magnetosphere could also lead to a larger overlap
between the radio and gamma-ray beams and explain the small fraction of radio-
quiet gamma-ray MSPs in the population.

2.4.2 Acceleration Gaps in the Magnetosphere

Compared to the other energy bands, the gamma-ray emission from pulsars is of
special importance since its luminosity is the only interesting fraction ($\sim$10% on
average) of the main spin-down power [3]. As discussed in Sect. 2.2, a force-free
configuration in the pulsar's magnetosphere implies $\mathbf{E} \cdot \mathbf{B} = 0$ and thus hinders high
energy emission. Particle acceleration in pulsars has therefore been studied primarily
on a local scale so far, where the force-free condition is dropped in certain acceleration zones or gaps. The formation of such gaps is based on the current outflow on
open magnetic field lines and hence the location of the gaps is limited to the open
magnetosphere, albeit at different altitudes depending on the model (see Fig. 2.14).
Three of the most studied model types, which mainly differ regarding the location
and geometry of the acceleration zone, are the polar cap (PC) model [53–55], the slot
gap (SG, also referred to as Two-Pole Caustic Model, TPC) model [51, 56, 57] and
the outer gap (OG) model [58–60].[7] In the polar cap model, the accelerating field $E_\parallel$
arises above the neutron star's polar caps up to a few stellar radii. The upper boundary
is formed by the so-called pair formation front (PFF) where the accelerated particles
are energetic enough to develop pair cascades, screening $E_\parallel$ within a relatively thin
region above the PFF [41]. Since $E_\parallel$ is assumed to vanish in the closed magnetosphere,
$E_\parallel$ is likely to decrease towards the last open field line. This requires the charges
to accelerate longer distances in order to obtain large enough Lorentz factors and to

---

[6]Although a strong hint at the 3.5 σ level was reported for the MSP PSR J0218+4232 [47].

[7]For an excellent assessment of the pulsar outer gap models, the reader is referred to [61].

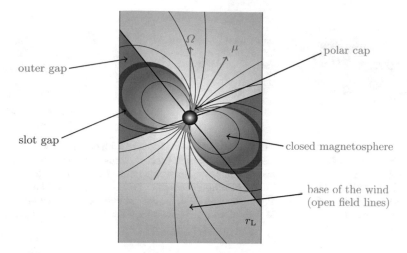

Fig. 2.14 A sketch of the assumed acceleration regions in a pulsar's magnetosphere. The rotation and dipole axes, Ω and μ, are denoted as green and red arrows, respectively. r_L denotes the light cylinder radius. All acceleration regions, or so-called *gaps* (see text), are located in the open magnetosphere. Figure taken from [52] (Color figure online)

radiate energetic enough photons needed for the pair cascades. Therefore, in the SG models the PFF curves upward along the last open field line forming a narrow slot gap that can extend all the way up to the light cylinder. In OG models the formation of the gap is based on a charge deficiency above the null charge surface, where ρ_{GJ} reverses sign and charges from the neutron star surface have difficulties to populate this region. Modern versions of the SG and OG models include also the production of pulsed gamma rays via inverse Compton scattering besides synchrocurvature radiation [51, 60]. Another type of emission models place the acceleration zone beyond the light cylinder in the current sheet of the pulsar wind [62, 63]. Interestingly, when leaving the local scale of acceleration gaps and turning to a more global scale, dissipative pulsar magnetosphere models exhibit accelerating electric fields in all of the gap regions mentioned above, as shown in Sect. 2.2.

The wealth of gamma-ray pulsars discovered by the *Fermi* LAT allowed population studies testing the different gamma-ray emission models [22, 42, 50, 64]. All of them favor or are in accordance with emission from the outer parts of the magnetosphere, either within or beyond the light cylinder. The recent advances in finding global configurations of the pulsar magnetosphere, actually seem to identify the current sheet in the pulsar wind or the region close to it as main source of gamma-ray photons [21, 23]. For models trying to explain the exceptional VHE emission from the Crab pulsar, the reader is referred to Sects. 4.3 and 5.6 of this thesis.

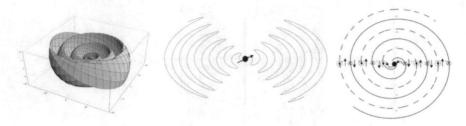

Fig. 2.15 Geometry of the so-called striped wind leaving the pulsar. **Left**: Geometry of one of the two spirally wound thin current sheets. **Middle**: Meridional cross section of the two current sheets for an oblique rotator of 60°. The sheets undulate within an angle around the equatorial plane tracing stripes. **Right**: Equatorial cross section showing the current sheets as solid and dashed line. The arrows show the local directions of the magnetic field, the dots and crosses show the direction of the current flow. The magnetic field lines reverse their direction at the current sheet. Figures taken from [65]

2.5 Evolution of Pulsar Wind Nebulae

As mentioned in the beginning of the chapter, most of the spin-down luminosity of a pulsar is carried away in form of a pulsar wind consisting of Poynting flux and relativistic particles. The ratio between the Poynting flux and the kinetic energy of the wind is expressed by the magnetisation or sigma-parameter σ. At the pulsar the wind is highly Poynting flux dominated ($\sigma \sim 10^3$) and forms current sheets that spiral outwards as illustrated in Fig. 2.15 [65]. Thus the energy reservoir for this striped wind is the neutron star's rotational energy. The time evolution of the energy output due to the spin-down of a pulsar can be written as[8]

$$\dot{E}(t) = \dot{E}_0 \left(1 + \frac{t}{\tau_0} \right)^{-\frac{n+1}{n-1}} , \qquad (2.11)$$

where $\dot{E}_0$ is the initial spin-down power, n the braking index of the system and τ_0 is the initial spin-down timescale defined as

$$\tau_0 \equiv \frac{P_0}{(n-1)\dot{P}_0} . \qquad (2.12)$$

τ_0 corresponds to the characteristic age defined in Eq. 2.8 at $t = 0$ and $n = 3$. The energy output is thus roughly constant until a time τ_0, beyond which $\dot{E} \propto t^{-(n+1)/(n-1)}$.

Compared to the energy of about $\sim 10^{51}$ erg contained in a supernova (SN) explosion shock, which drives the expansion of the supernova remnant (SNR) in the interstellar medium (ISM), the spin-down power of a pulsar is negligible. Therefore, the dynamical evolution of the SNR will hardly be affected by the energy output of the pulsar, whereas the evolution of the pulsar and its system strongly depends on the

---

[8]For a comprehensive derivation of the formula, see Appendix B in [4].

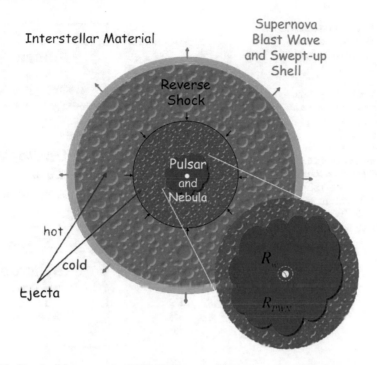

Fig. 2.16 Sketch of the composite PWN-SNR system. The inset magnifies the PWN region, where R_w denotes the termination shock (white dotted circle) and R_{pwn} the front shock of the synchrotron nebula. The sketch corresponds to the free expanding stage of a PWN, since the front shock of the PWN still has to interact with the reverse shock of the SNR. Figure taken from [66]

interaction with the SN explosion and remnant. When electromagnetic radiation from the PWN and a surrounding SN shell is observed, the system is termed *composite* (see Fig. 2.16). The time evolution of a PWN can generally be divided into three stages: the free expanding ($< 2 - 6$ kyr), the reverse shock interaction (until some tens of kyr), and the relic stage.

After its birth a pulsar is embedded in the cold ejecta from the SN explosion and has a constant energy output (see Eq. 2.11, $t \ll \tau_0$). The energy output of the pulsar in form of a cold pulsar wind[9] is concentrated towards the spin equator and develops a so-called termination shock when it first interacts with the SN ejecta. The termination shock of the wind forms at a radius where the ram pressure of the wind is balanced by the thermal pressure of the PWN. Due to the anisotropy of the wind power, the shock will reach pressure equilibrium at different locations often resulting in a torus-like structure, as for example in the Crab nebula as shown on the right of Fig. 2.18 [68]. Beyond this termination shock a pulsar wind nebula (PWN) builds up as an expanding bubble of diffuse plasma confined by the expanding shell of the supernova remnant (SNR). At the termination shock, particles are thermalized and reaccelerated, possibly

---

[9]Here the adjective "cold" means that the wind's thermal energy is much smaller than its magnetic and bulk kinetic energy.

Fig. 2.17 Radiation zones in a pulsar wind nebula. The scale is similar to the inset of Fig. 2.16, except the magnified unshocked wind zone. While the pulsar and the synchrotron nebula emit photons in a wide energy range (from radio to gamma-rays, mainly via synchrotron radiation and inverse Compton scattering), the unshocked wind is thought to be able to produce only GeV or TeV photons. Figure taken from [67]

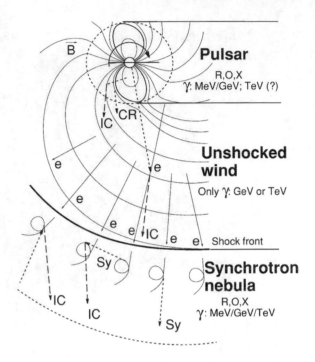

via the *Fermi* acceleration mechanism [69, 70] or by magnetic reconnection of the striped wind [71]. Downstream of the flow, the accelerated particles emit synchrotron radiation from which the synchrotron nebula emerges (see Fig. 2.17). Upstream of the termination shock the magnetisation parameter σ is believed to be $\sigma \simeq 10^{-3} - 10^{-2}$ (inferred from 1D and 2D PWN models, see for example [72]), quite in contrast to the value when the wind leaves the pulsar. This discrepancy is known as the σ-problem and is still an very active research topic. Solutions try to either convert magnetic energy to kinetic energy in the unshocked wind or find PWN models that work with a high σ (for an extensive discussion of the σ-problem, [68] see and references therein).

As the SNR expands freely into the ISM, the PWN is over-pressured with respect to its environment and expands supersonically driving a shock into the cold SN ejecta at R_{pwn} (see Fig. 2.16). The cold ejecta is much denser than the PWN bubble and chaotic, filamentary structures emerge via *Rayleigh-Taylor* instabilities in the contact discontinuity between the PWN and the SNR [73]. Well known examples of these Rayleigh-Taylor filaments are found in the Crab nebula, as shown in the left of Fig. 2.18.

When the SNR reaches the end of its free expansion stage, a reverse shock is driven towards the interior of the system, which marks the beginning of the *Sedov-Taylor* phase and heats up the stellar ejecta [74]. The collision of the SNR reverse shock with the PWN forward shock can crush the PWN and reduce R_{pwn} by a large factor [66]. Once the reverse shock of the SN is stopped, the PWN bubble can continue to grow steadily and expands now into the hot shocked SN ejecta at subsonic speed.

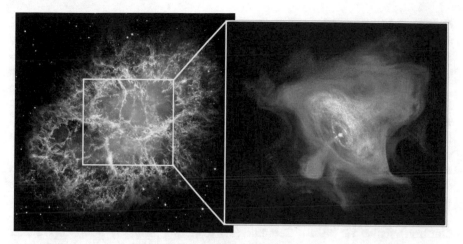

Fig. 2.18 The Rayleigh-Taylor (RT) instabilities and the torus structure of the Crab nebula. **Left**: The optical image of the Crab nebula taken by the Hubble Space telescope illustrates the RT instabilities (filamentary, finger-like structures) that emerge in the contact discontinuity between the PWN and the cold ejecta of the SN explosion (see Fig. 2.16). Image taken from http://hubblesite.org/ image/1823/news/79-pulsars, last accessed 21/09/2017. **Right**: In X-rays the torus-like structure of the termination shock is visible as an inner ring around the pulsar (white dot in the middle). Image taken from http://chandra.harvard.edu/photo/2008/crab/, last accessed 21/09/2017

In the time scales of the Sedov phase, the pulsar, if it was given a significant kick during the SN explosion, is able to travel distances comparable to or even larger than the radius of its PWN. Typical proper velocities of pulsars are in the range of 400–500 km s^{-1}, but can reach over 1000 km s^{-1}s. The pulsar thus escapes from its original wind bubble and leaves behind a so-called relic PWN, while powering a new, smaller PWN at its current position. If the pulsar moves supersonically through its new environment, a ballistic bow shock is produced that tightly confines the new PWN to a rather small extension of $\lesssim 1$ pc [66]. Such bow-shock PWNe take on a cometary appearance when observed at X-ray and radio energies. Due to their high proper velocities, pulsars are able to move far away from their birthplace into the ISM and even escape the galaxy, which explains the population of old MSPs found outside the Milky Way. But long before this escape, the energy output will eventually become insufficient to power an observable synchrotron nebula and the pulsar will be surrounded by a ghost nebula.

2.6 TeV Emission of Pulsar Wind Nebulae

To date around $\sim$100 pulsar wind nebulae are known, most of them initially detected in X-rays by the Chandra X-ray Observatory [75]. Around $\sim$50 of them have VHE associations or possible counterparts, making PWNe the largest population among

Galactic TeV sources. The association of a PWN to a VHE source, however, is often a difficult and complex task due to the large extension of TeV PWNe and the relatively poor angular and energy resolution of current VHE instruments. Theoretically one expects that especially powerful and young pulsars grow bright TeV PWNe, since most of the pulsar's spin-down energy is deployed within few tens of kiloyears (see Eq. 2.12). A way to test this association experimentally was explored by Carrigan et al. [76] who used the second release of the H.E.S.S. Galactic Plane Survey (HGPS, [77]) and the Parks Multibeam Pulsar Survey (PMPS, [78]) to investigate the fraction of VHE detections around pulsars (within 0.22 deg) to the total number of pulsars in different $\dot{E}/d^2$ bands, where d denotes the distance to the pulsar. This positive spatial correlation was recently affirmed by Abdalla et al. [4] with an updated version of the HGPS.

The production of GeV to TeV photons is possible in the unshocked and synchrotron nebulae via inverse Compton (IC) scattering primarily on the interstellar radiation field (ISRF) and the CMB. In young PWNe like the Crab nebula, the synchrotron self-Compton (SSC) component from the synchrotron nebula can dominate the luminosity at VHE (see Fig. 2.19). Even though the IC component of a PWN's spectrum is energetically subordinated, it carries unique information about the electron population in the nebula that the synchrotron component does not give access to [4]. Since the IC photons emerge from photon seed fields (ISRF and CMB) that are homogeneous and constant in time, they trace the electron plasma more accurately than the synchrotron photons, which depend on the space- and time-varying magnetic fields in the PWN. Additionally, the low energy part of the electron spectrum yields a longer synchrotron cooling time and is able to diffuse farther away from the pulsar, where it can only be detected by its IC emission. Hence, TeV PWNe are in general far more extended than their X-ray counterparts and give a more complete picture of the underlying electron population.

Using the current nine-year HGPS catalog and observations made by VERITAS and MAGIC, Abdalla et al. [4] conducted an extensive population study based primarily on the 19 firmly identified PWNe that have been observed by IACTs so far. Furthermore, they were able to identify 20 additional candidates, out of which 10 are considered to be viable TeV PWNe. Trying to minimize observational biases in their study, they also included upper limits from regions around pulsars with $\dot{E} > 10^{35}$ erg s^{-1}, derived from the HGPS. For the first time they could provide a significant numerical context for the correlation between luminosity[10] $L_{1-10\text{TeV}}$ and spin-down power $\dot{E}$, and estimated this power-law relation to be $L_{1-10\text{TeV}} \propto \dot{E}^{0.58\pm0.21}$. This correlation was implicitly known before by the mere fact of missing TeV PWN detections around old pulsars, but a numerical description was still missing. While in X-rays a clear correlation between luminosity and spin-down power, as well as characteristic age τ_c is seen (see Fig. 2.20), in TeV the latter correlation is not so obvious (p-value $= 0.13$). On the other hand [4] were able to confirm the expansion of the nebula with time ($R_{\text{pwn}} \propto \tau^{0.55\pm0.23}$), which is not so clear in X-rays since the synchrotron cooling time is too short for the electrons to reach the outer part of the nebula. A

---

[10]$L_{1-10\text{TeV}}$ denotes the luminosity integrated from 1 to 10 TeV.

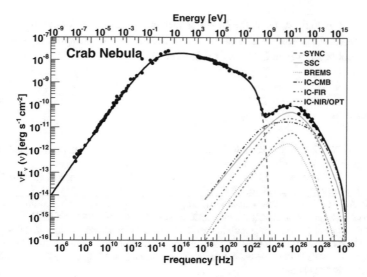

Fig. 2.19 The broadband spectrum of the Crab nebula with the fitted time dependent model by [79]. While the synchrotron (SYNC) component covers the emission from radio to X-rays, the TeV emission is explained mainly by IC scattering. The interstellar radiation field (ISRF) is divided into its far infrared (FIR) and near infrared (NIR)/optical (OPT) components. The model also takes into account the emission from Bremsstrahlung (BREMS). In young PWNe like the Crab nebula, the synchrotron self-Compton (SSC) component can be the main contributer in the VHE range. Plot taken from [80] (Color figure online)

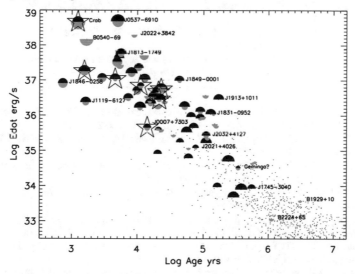

Fig. 2.20 Pulsars and their detected PWNe (or PWN candidates) plotted in the $\dot{E} - \tau_c$ space. The semi-circles correspond to X-ray (orange) and TeV (black) PWNe, while their sizes are proportional to the logarithms of their luminosities. Stars denote the PWNe that are detected by the *Fermi* LAT at GeV energies. The small black dots depict pulsars from the ATNF catalog [24]. While in X-rays the correlation between luminosity and $\dot{E}$ (τ_c, respectively) is obvious, at TeV energies the latter correlation is not so clear. Plot taken from [75] (Color figure online)

shortcoming of their study is the disregard of the photon seed fields at the position of each pulsar. This issue will partially be addressed by a forthcoming paper by the MAGIC collaboration (see Chap. 8 of this thesis).

References

1. Fryer CL, New KCB (2011) Gravitational waves from gravitational collapse. Living Rev Relativ 14(1):1. https://doi.org/10.12942/lrr-2011-1
2. Lattimer JM, Prakash M (2004) The physics of neutron stars. Science 304(5670):536–542. https://doi.org/10.1126/science.1090720
3. Abdo AA et al (2013) The second fermi large area telescope catalog of gamma-ray pulsars. Astrophys J Suppl Ser 208(2):17. https://doi.org/10.1088/0067-0049/208/2/17
4. Abdalla H et al (2018) The population of TeV pulsar wind nebulae in the H.E.S.S. Galactic Plane Survey. Astron Astrophys 612:A2. https://doi.org/10.1051/0004-6361/201629377
5. Chadwick J (1932) Possible existence of a neutron. Nature 129(3252):312–312. https://doi.org/10.1038/129312a0
6. Baade W, Zwicky F (1934) Remarks on super-novae and cosmic rays. Phys Rev 46(1):76–77. https://doi.org/10.1103/PhysRev.46.76.2
7. Hewish A et al (1968) Observation of a rapidly pulsating radio source. Nature 217(5130):709–713. https://doi.org/10.1038/217709a0
8. Page D, Reddy S (2006) Dense matter in compact stars: theoretical developments and observational constraints. Annu Rev Nucl Part Sci 56(1):327–374. https://doi.org/10.1146/annurev.nucl.56.080805.140600
9. Harding AK (2013) The neutron star zoo. Front Phys 8(6):679–692. https://doi.org/10.1007/s11467-013-0285-0
10. Archibald AM et al (2009) A radio pulsar/X-ray binary link. Science 324(5933):1411–1414. https://doi.org/10.1126/science.1172740
11. Papitto A et al (2013) Swings between rotation and accretion power in a binary millisecond pulsar. Nature 501(7468):517–520. https://doi.org/10.1038/nature12470
12. Goldreich P, Julian WH (1969) Pulsar Electrodynamics. Astrophys J 157:869. https://doi.org/10.1086/150119
13. Timokhin AN (2006) On the force-free magnetosphere of an aligned rotator. Mon Not R Astron Soc 368(3):1055–1072. https://doi.org/10.1111/j.1365-2966.2006.10192.x
14. Michel FC (1973) Rotating magnetospheres: an exact 3-D solution. Astrophys J 180:L133. https://doi.org/10.1086/181169
15. Scharlemann ET, Wagoner RV (1973) Aligned rotating magnetospheres. General analysis. Astrophys J 182:951. https://doi.org/10.1086/152195
16. Contopoulos I et al (1999) The axisymmetric pulsar magnetosphere. Astrophys J 511(1):351–358. https://doi.org/10.1086/306652
17. Spitkovsky A (2006) Time-dependent force-free pulsar magnetospheres: axisymmetric and oblique rotators. Astrophys J 648(1):L51–L54. https://doi.org/10.1086/507518
18. Gruzinov A (2007) Pulsar emission and force-free electrodynamics. Astrophys J 667(1):L69–L71. https://doi.org/10.1086/519839
19. Li J et al (2012) Resisitive solutions for pulsar magnetospheres. Astrophys J 746(1):60. https://doi.org/10.1088/0004-637X/746/1/60
20. Kalapotharakos C et al (2012) Toward a realistic pulsar magnetosphere. Astrophys J 749(1):2. https://doi.org/10.1088/0004-637X/749/1/2
21. Kalapotharakos C et al (2014) Gamma-ray emission in dissipative pulsar magnetospheres: from theory to fermi observations. Astrophys J 793(2):97. https://doi.org/10.1088/0004-637X/793/2/97

22. Brambilla G et al (2015) Testing dissipative magnetosphere model light curves and spectra with Fermi pulsars. Astrophys J 804(2):84. https://doi.org/10.1088/0004-637X/804/2/84

23. Philippov AA et al (2015) Ab initio pulsar magnetosphere: three-dimensional particle-in-cell simulations of oblique pulsars. Astrophys J 801(1):L19. https://doi.org/10.1088/2041-8205/801/1/L19

24. Manchester RN et al (2005) The Australia telescope national facility pulsar catalogue. Astron J 129(4):1993–2006. https://doi.org/10.1086/428488

25. Michel F, Li H (1999) Electrodynamics of neutron stars. Phys Rep 318(6):227–297. https://doi.org/10.1016/S0370-1573(99)00002-2

26. Shapiro SL, Teukolsky SA (1983) Black holes. The physics of compact objects. Wiley-VCH, White Dwarfs and Neutron Stars. ISBN 0471873160

27. Jackson JD (1998) Classical electrodynamics, 3rd edn. Wiley. ISBN 047130932X

28. Gunn JE, Ostriker JP (1969) Magnetic dipole radiation from pulsars. Nature 221:454. https://doi.org/10.1038/221454a0

29. Tong H, Kou FF (2017) Possible evolution of the pulsar braking index from larger than three to about one. Astrophys J 837(2):117. https://doi.org/10.3847/1538-4357/aa60c6

30. de Araujo JC et al (2016) Gravitational wave emission by the high braking index pulsar PSR J1640–4631. J Cosmol Astropart Phys 2016(07):023. https://doi.org/10.1088/1475-7516/2016/07/023

31. Mignani RP (2011) Optical, ultraviolet, and infrared observations of isolated neutron stars. Adv Space Res 47(8):1281–1293. https://doi.org/10.1016/j.asr.2009.12.011

32. Mignani RP et al (2016) Observations of three young γ-ray pulsars with the Gran Telescopio Canarias. Mon Not R Astron Soc 461(4):4317–4328. https://doi.org/10.1093/mnras/stw1629

33. Becker W (2009) Neutron stars and pulsars, vol 357 of Astrophysics and space science library. Springer, Berlin. https://doi.org/10.1007/978-3-540-76965-1. ISBN 978-3-540-76964-4

34. Ackermann M et al (2012) The Fermi large area telescope on orbit: event classification, instrument response functions, and calibration. Astrophys J Suppl Ser 203(1):4. https://doi.org/10.1088/0067-0049/203/1/4

35. Thompson DJ (2004) Gamma ray pulsars. In: Cheng KS, Romero GE (eds) Cosmic gamma-ray sources. Kluwer Academic, Dordrecht, pp 149–168. https://doi.org/10.1007/978-1-4020-2256-2_7

36. Thompson DJ (2008) Gamma ray astrophysics: the EGRET results. Rep Prog Phys 71(11):116901. https://doi.org/10.1088/0034-4885/71/11/116901

37. Baring MG (2004) High-energy emission from pulsars: the polar cap scenario. Adv Space Res 33(4):552–560. https://doi.org/10.1016/j.asr.2003.08.020

38. Lee KJ et al (2010) Low bounds for pulsar γ-ray radiation altitudes. Mon Not R Astron Soc 405(3):2103–2112. https://doi.org/10.1111/j.1365-2966.2010.16600.x

39. Watters KP, Romani RW (2011) The galactic population of young γ-ray pulsars. Astrophys J 727(2):123. https://doi.org/10.1088/0004-637X/727/2/123

40. Watters KP et al (2009) An atlas for interpreting γ-ray pulsar light curves. Astrophys J 695(2):1289–1301. https://doi.org/10.1088/0004-637X/695/2/1289

41. Grenier IA, Harding AK (2006) Pulsar twinkling and relativity. In: AIP conference proceedings, p 9

42. Viganò D et al (2015) A systematic synchro curvature modelling of pulsar γ-ray spectra unveils hidden trends. Mon Not R Astron Soc 453(3):2600–2622. https://doi.org/10.1093/mnras/stv1582

43. Bochenek C, McCann A (2015) On the spectral shape of gamma-ray pulsars above the break energy. In: 34th international cosmic ray conference. The Hague, Netherlands

44. Ahnen ML et al (2016) Search for VHE gamma-ray emission from Geminga pulsar and nebula with the MAGIC telescopes. Astron Astrophys 591:A138. https://doi.org/10.1051/0004-6361/201527722

45. Abdo AA et al (2010c) The Vela pulsar: results from the first year of Fermi LAT observations. Astrophys J 713(1):154–165. https://doi.org/10.1088/0004-637X/713/1/154

46. Takata J et al (2016) Probing gamma-ray emissions of Fermi -LAT pulsars with a non-stationary outer gap model. Mon Not R Astron Soc 455(4):4249–4266. https://doi.org/10.1093/mnras/stv2612

47. Kuiper L et al (2000) The likely detection of pulsed high-energy γ-ray emission from millisecond pulsar PSR J0218+4232. Astron Astrophys 359:615–626

48. Harding AK et al (2002) Regimes of pulsar pair formation and particle energetics. Astrophys J 576(1):366–375. https://doi.org/10.1086/341633

49. Harding AK et al (2005) High energy emission from millisecond pulsars. Astrophys J 622(1):531–543. https://doi.org/10.1086/427840

50. Johnson TJ et al (2014) Constraints on the emission geometries and spin evolution of gamma-ray millisecond pulsars. Astrophys J Suppl Ser 213(1):6. https://doi.org/10.1088/0067-0049/213/1/6

51. Harding AK, Kalapotharakos C (2015) Synchrotron self-compton emission from the Crab and other pulsars. Astrophys J 811(1):63. https://doi.org/10.1088/0004-637X/811/1/63

52. Pétri J (2016) Theory of pulsar magnetosphere and wind. J Plasma Phys 82(05):635820502. https://doi.org/10.1017/S0022377816000763

53. Sturrock PA (1971) A model of pulsars. Astrophys J 164:529. https://doi.org/10.1086/150865

54. Ruderman MA, Sutherland PG (1975) Theory of pulsars—polar caps, sparks, and coherent microwave radiation. Astrophys J 196:51. https://doi.org/10.1086/153393

55. Daugherty JK, Harding AK (1996) Gamma-ray pulsars: emission from extended polar CAP cascades. Astrophys J 458:278

56. Arons J (1983) Pair creation above pulsar polar caps—geometrical structure and energetics of slot gaps. Astrophys J 266:215. https://doi.org/10.1086/160771

57. Dyks J, Rudak B (2003) Two pole caustic model for high energy light curves of pulsars. Astrophys J 598(2):1201–1206. https://doi.org/10.1086/379052

58. Cheng KS et al (1986) Energetic radiation from rapidly spinning pulsars. I-Outer magnetosphere gaps. II–VELA and Crab. Astrophys J 300:500. https://doi.org/10.1086/163829

59. Romani RW (1996) Gamma-ray pulsars: radiation processes in the outer magnetosphere. Astrophys J 470:469. https://doi.org/10.1086/177878

60. Hirotani K (2015) Three-dimensional non-vacuum pulsar outer-gap model: localized acceleration electric field in the higher altitudes. Astrophys J 798(2):L40. https://doi.org/10.1088/2041-8205/798/2/L40

61. Vigano D et al (2015a) An assessment of the pulsar outer gap model—I. Assumptions, uncertainties, and implications on the gap size and the accelerating field. Mon Not R Astron Soc 447(3):2631–2648. https://doi.org/10.1093/mnras/stu2564

62. Pétri J (2012) High-energy emission from the pulsar striped wind: a synchrotron model for gamma-ray pulsars. Mon Not R Astron Soc 424(3):2023–2027. https://doi.org/10.1111/j.1365-2966.2012.21350.x

63. Arka I, Dubus G (2013) Pulsed high-energy γ-rays from thermal populations in the current sheets of pulsar winds. Astron Astrophys 550:A101. https://doi.org/10.1051/0004-6361/201220110

64. Pierbattista M et al (2012) Constraining γ-ray pulsar gap models with a simulated pulsar population. Astron Astrophys 545:A42. https://doi.org/10.1051/0004-6361/201219135

65. Arons J (2012) Pulsar wind nebulae as cosmic pevatrons: a current sheet's tale. Space Sci Rev 173(1–4):341–367. https://doi.org/10.1007/s11214-012-9885-1

66. Gaensler BM, Slane PO (2006) The evolution and structure of pulsar wind nebulae. Annu Rev Astron Astrophys 44(1):17–47. https://doi.org/10.1146/annurev.astro.44.051905.092528

67. Aharonian F, Bogovalov S (2003) Exploring physics of rotation powered pulsars with sub-10 GeV imaging atmospheric Cherenkov telescopes. New Astron 8(2):85–103. https://doi.org/10.1016/S1384-1076(02)00200-2

68. Porth O et al (2017) Modelling jets, tori and flares in pulsar wind nebulae. Space Sci Rev 207(1–4):137–174. https://doi.org/10.1007/s11214-017-0344-x

69. Blandford R, Eichler D (1987) Particle acceleration at astrophysical shocks: a theory of cosmic ray origin. Phys Rep 154(1):1–75. https://doi.org/10.1016/0370-1573(87)90134-7

70. Spitkovsky A (2008) Particle acceleration in relativistic collisionless shocks: Fermi process at last? Astrophys J 682(1):L5–L8. https://doi.org/10.1086/590248
71. Sironi L, Spitkovsky A (2011) Acceleration of particles at the termination shock of a relativistic striped wind. Astrophys J 741(1):39. https://doi.org/10.1088/0004-637X/741/1/39
72. Kennel CF, Coroniti FV (1984) Confinement of the Crab pulsar's wind by its supernova remnant. Astrophys J 283:694. https://doi.org/10.1086/162356
73. Porth O et al (2014) Rayleigh-Taylor instability in magnetohydrodynamic simulations of the Crab nebula. Mon Not R Astron Soc 443(1):547–558. https://doi.org/10.1093/mnras/stu1082
74. van der Swaluw E et al (2001) Pulsar wind nebulae in supernova remnants. Astron Astrophys 380(1):309–317. https://doi.org/10.1051/0004-6361:20011437
75. Kargaltsev O et al (2013) Gamma-ray and X-ray properties of pulsar wind nebulae and unidentified Galactic TeV sources, p 16
76. Carrigan S et al (2007) Establishing a connection between high-power pulsars and very-high-energy gamma-ray sources. In: 30th international cosmic ray conference, Mcirda, Mexico, pp 659–662
77. Aharonian F et al (2006) The H.E.S.S. survey of the inner galaxy in very high energy gamma rays. Astrophys J 636(2):777–797. https://doi.org/10.1086/498013
78. Lorimer DR et al (2006) The Parkes Multibeam Pulsar Survey VI. Discovery and timing of 142 pulsars and a Galactic population analysis. Mon Not R Astron Soc 372:777–800. https://doi.org/10.1111/j.1365-2966.2006.10887.x
79. Martin J et al (2012) Time-dependent modelling of pulsar wind nebulae: study on the impact of the diffusion-loss approximations. Mon Not R Astron Soc 427(1):415–427. https://doi.org/10.1111/j.1365-2966.2012.22014.x
80. Torres D et al (2014) Time-dependent modeling of TeV-detected, young pulsar wind nebulae. J High Energy Astrophys 1–2:31–62. https://doi.org/10.1016/j.jheap.2014.02.001

Chapter 3
Cherenkov Telescopes and MAGIC

In the Ph.D. dissertations of students studying the atmospheric Cherenkov phenomenon the first reference should be to the 1948 note by the British Nobel Laureate, P.M.S. Blackett in the Royal Society report on the study of night-sky light and aurora.

Trevor C. Weekes, 2005

The first source of very-high-energy (VHE) gamma rays, the Crab nebula, was firmly detected by the Whipple collaboration in the 1980s [1]. Their pioneering instrument and analysis method became the basic concept in the field of Imaging Atmospheric Cherenkov Telescopes (IACTs) and paved the way for todays third generation telescopes. The Major Atmospheric Gamma-ray Imaging Cherenkov (MAGIC) telescopes, located at the Roque de los Muchachos Observatory in La Palma, Canary Islands (Spain), are one of the three major IACTs currently operating. Together with the other two IACTs, H.E.S.S. and VERITAS, they contributed to the breakthrough of the IACT technique and proved the technique to be the most sensitive in the GeV to TeV energy range. Around 90% of the VHE sources listed in the TeVCat[1] were discovered by the current three major IACTs. Since the pioneering days, many hardware improvements, as well as advances in the analysis methods, have helped to lower the energy threshold of IACTs down to $\sim$30 GeV and have enabled IACTs to detect VHE photon fluxes of $\sim$0.5% of the Crab nebula flux in less than 50 h.

In the first two sections of this chapter we will give a short overview to the IACT technique, before introducing in more detail the MAGIC telescopes in Sect. 3.3. The chapter concludes with Sect. 3.4, in which we discuss the analysis of MAGIC data.

---

[1] Wakely and Horan [2], http://tevcat.uchicago.edu, last accessed 25/09/2017.

© Springer Nature Switzerland AG 2019
D. Carreto Fidalgo, *Revealing the Most Energetic Light from Pulsars and Their Nebulae*, Springer Theses, https://doi.org/10.1007/978-3-030-24194-0_3

3.1 Air Showers and Atmospheric Cherenkov Radiation

As mentioned in Chap. 1 and illustrated in Fig. 1.1, the Earth's atmosphere is non-transparent for gamma rays. Two physical phenomena that nevertheless allow for ground-based astronomy in the VHE band, are Extensive Air Showers (EAS) and Cherenkov radiation.

When entering the Earth's atmosphere, cosmic rays, as well as very-high-energy gamma rays, can interact with the present molecules through various processes leading to the development of a particle shower called Extensive Air Shower (EAS). In the case of gamma rays, the first interaction takes place at around 20–30 km above sea level (a.s.l.) depending on the energy of the photon. The EAS started by a primary gamma-ray evolves from an electromagnetic cascade due to the interplay of two processes: (i) pair production, which converts a high-energy photon in an electron-positron pair ($e^{\pm}$) in the Coulomb field of an atmospheric nuclei; (ii) bremsstrahlung emission from $e^{\pm}$ in the same Coulomb field leading to the production of further high-energy gamma rays. The material in which the shower develops is characterized by its *radiation length* X_0, measured in g cm^{-2}. This characteristic parameter indicates the mean amount of matter, projected on a plane, that an electron must traverse to loose $1/e$ of its energy by bremsstrahlung. At the same time the mean free path for the pair production is $\frac{7}{9}X_0$ [3]. The similar length scale for both interactions leads to a compact and symmetric shower structure about the main shower axis as can be observed in Fig. 3.2. Dry air, as found in the atmosphere, has a radiation length of 36.7 g cm^{-2} [4]. This makes the atmosphere a relatively thick calorimeter

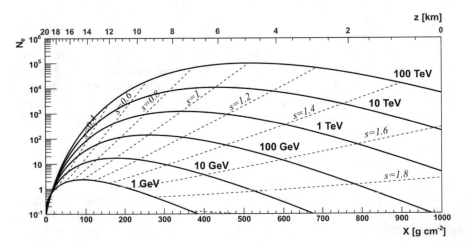

Fig. 3.1 Longitudinal development of an Extensive Air Shower initiated by an gamma ray. The number of secondary electrons is plotted for various photon energies versus the radiation length (lower x-axis) and the height above sea level (upper x-axis). The diagonal dashed lines denote the shower age s, for which $s = 0$ is the start of the shower, $s = 1$ at the shower maximum and $s > 1$ corresponds to the following extinction phase. Figure taken from [4]

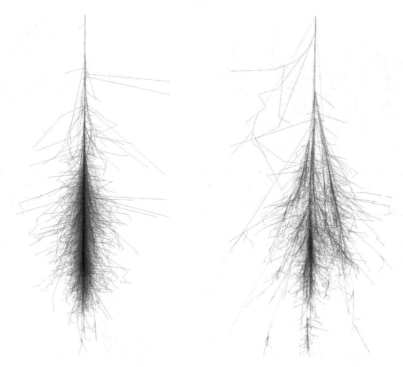

Fig. 3.2   Particle tracks of two simulated Extensive Air Showers (EAS). The left shower corresponds to an electromagnetic cascade initiated by a gamma ray, the right shower illustrates an hadronic shower. Both showers were simulated for a primary particle with an energy of 100 GeV. Images taken from https://www.ikp.kit.edu/corsika/, last accessed 27/09/2017

of approximately $\sim$27 X_0, compared to the $\sim$9 X_0 that are used for calorimeters in gamma-ray satellites such as *Fermi*-LAT [5]. However, the atmosphere is a strongly inhomogeneous calorimeter, in which at sea level one radiation length corresponds to $\sim$300 m, whereas at 10 km it is $\sim$1 km. As a consequence the shower maximum h_{max}, defined as height above sea level where the number of shower particles is at its maximum, depends weakly on the energy of the primary particle E_0, that is $h_{\mathrm{max}} \propto \log(\log(E_0))$ instead of $h_{\mathrm{max}} \propto \log(E_0)$ as is the case for an homogeneous calorimeter. For a primary photon energy between 50 GeV and 10 TeV, h_{max} is around 10–6 km (see Fig. 3.1). After that the shower dies out because the energy loss due to ionization processes becomes dominant over the loss via pair creation.

Extensive Air Showers initiated by hadrons exhibit a more diffuse structure compared to the air showers from gamma rays, as seen on the right in Fig. 3.2. The primary particle (usually a proton) collides with a nucleus in the atmosphere creating mostly pions (90% of the secondary particles are pions). Neutral pions π_0 have a very short life time ($\sim$10^{17} s) and decay into two photons quasi instantly. These high energy photons are able to induce electromagnetic cascades forming sub-showers in the EAS. Charged pions $\pi_{\pm}$, however, have a life time of 26 ns and can interact

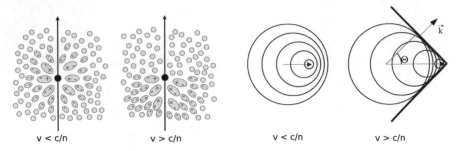

Fig. 3.3 Sketch of the Cherenkov light mechanism. **Left**: A charged particle polarizes the traversing medium. **Right**: The coherent depolarization of the medium in the case of $v > c/n$ results in a forward-beamed emission, called Cherenkov radiation. It is emitted along a cone with opening angle Θ, also known as Cherenkov angle. Illustrations taken from [4]

with further nuclei of the atmosphere, before decaying into a muon $\mu_\pm$ and a neutrino (antineutrino). Owing to relativistic effects, muons are able to reach the Earth's surface before decaying. In general, hadronic EAS are more complicated to describe and their various components do not exhibit a common length scale as is the case for electromagnetic cascades.

The second phenomena, that enables ground-based VHE astronomy, is Cherenkov radiation[2] [6]. Cherenkov radiation is emitted when a relativistic charged particle moves through a medium with a speed $v > c/n$, where c is the vacuum speed of light and n is the refraction index of the medium. The charged particle polarizes the medium, which emits spherical electromagnetic waves along the track of the relativistic particle (see Fig. 3.3). The fact that the particle moves faster than the electromagnetic waves, leads to a positive interference at an angle Θ_c, called the Cherenkov angle [7]. Hence, the radiation will be emitted in a cone at the angle Θ_c and cancels out in all other directions as illustrated in the right of Fig. 3.3. If we define the ratio between the speed of the particle and the speed of light in vacuum as $\beta = v/c$, the Cherenkov angle is obtained by $\cos(\theta_c) = 1/n\beta$. In air, θ_c is about 1.5° at sea level, decreasing quasi linearly to $\sim$0.2° at $\sim$30 km [4]. The energy threshold at which the particle is able to emit Cherenkov radiation depends on the refraction index n and increases with height.

On the Earth's surface the Cherenkov light of each particle track illuminates a donut shaped ring that adds up to form the so-called Cherenkov light pool created by the whole shower as shown on the left of Fig. 3.4. The photon density of an air shower induced by a gamma-ray is roughly constant in a radius of $\sim$120 m centered on the shower core at 2200 m a.s.l. At $\sim$120 m a little hump in the density profile is observed which is a consequence of the opening of θ_C as particles penetrate deeper into the atmosphere. After this hump the photon density decreases rapidly. For a 100 GeV gamma-ray induced shower the photon density in the core of the Cherenkov light pool is $\sim$15 ph/m^2 at 2200 m a.s.l., as shown on the right of Fig. 3.4. The

---

[2]Strictly speaking, this phenomena just serves to detect or trace the particles of the shower and can be replaced by other physical processes, such as scintillation.

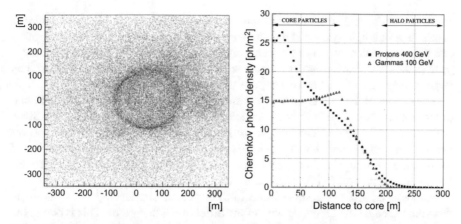

Fig. 3.4 A simulated Cherenkov light pool. **Left**: The Cherenkov photon distribution is shown for a 50 GeV gamma-ray induced EAS. At 2200 m a.s.l. (corresponding to the altitude of the MAGIC telescopes, see Sect. 3.3) the Cherenkov light typically illuminates a circle with a radius of ∼120 m. For an inclined shower the circle becomes elliptic. Figure taken from [8]. **Right**: Radial distribution of the photon density for a simulated gamma-ray shower and a proton-induced shower. Figure taken from [9]

photon density is proportional (first order) to the energy of the primary gamma ray as discussed further down. In general, hadronic showers exhibit a wider spread of their Cherenkov photons compared to gamma-ray induced EAS, which ultimately helps to distinguish both types of showers.

The spectral intensity of Cherenkov radiation is proportional to λ^{-2}, λ being the wavelength of the emitted radiation [7]. As seen on the left of Fig. 3.5, most of the light is emitted in the UV range, but absorption in the atmosphere shifts the spectral

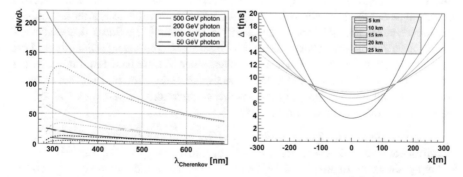

Fig. 3.5 Spectrum of the Cherenkov light of an air shower and its time distribution at the ground. **Left**: Spectra of the Cherenkov radiation for different energies of the primary gamma-ray. The dashed lines take into account absorption mechanisms, Rayleigh scattering and Mie scattering. Figure taken from [10]. **Right**: Plotted is the time delay Δt of the Cherenkov photons on the ground versus the lateral distance from the shower axis x for various altitudes of emission. Figure taken from [4] (Color figure online)

maximum of the detected photons on the ground to larger wavelengths. Transmission
losses in air are mainly caused by Rayleigh scattering off air molecules, absorption
due to ozone molecules and Mie scattering off aerosols. These processes produce a
peak in the Cherenkov photon spectrum at roughly $\lambda \simeq 320$ nm with a shift towards
larger wavelengths for an increasing zenith angle of the air showers.

The energy of the primary particle initiating an electromagnetic shower is in first
order proportional to the summed track length of all charged shower particles. Since
the number of emitted Cherenkov photons is proportional to the number of shower
particles, the total Cherenkov light yield is proportional in first order to the energy of
the primary particle. This fact is important when reconstructing the primary particles
energy from a recorded air shower on Earth.

Although an EAS needs several tens of microseconds to develop, the Cherenkov
light flash seen on the ground is only of the order of ~ 2 to ~ 10 ns. This is due to the
fact that the Cherenkov light radiated in the beginning of the shower development
is slower than the speed of the particles, but Cherenkov photons emitted closer to
the ground have in general a longer way to the observer (including the tracks of
the parent particles). These two effects can cancel out the development time of the
shower (see Fig. 3.5).

For an extensive discussion of the two phenomena, EAS and Cherenkov radiation,
and its application to the IACT technique, the reader is referred to [4] and references
therein.

3.2   Atmospheric Cherenkov Telescopes and Image Parameters

In 1953, Jelley and Galbraith [11] were the first to observe Cherenkov light from
EAS using a single photomultiplier tube (PMT). Still today PMTs are the preferred
choice to detect the faint nanosecond light flashes in the night sky, although silicon
photomultipliers are now on the brink of taking over [12]. Since this discovery
the setup to detect atmospheric Cherenkov light has improved significantly and the
shower detection rate went from a few events per minute to a few hundreds per
second. To achieve this rate and at the same time to reduce the background as much
as possible, all modern Imaging Atmospheric Cherenkov Telescopes (IACTs) are
based on the following four concepts [4]:

- To collect as much light as possible, especially from low-energy air showers, large
  mirrors are needed. Therefore, IACTs with a low energy threshold (~ 100 GeV)
  have dishes with diameters of 12–28 m, segmented in individual smaller mirrors
  of ~ 1 m^2 for technical and cost reasons.
- A spatially resolved image of the EAS is needed to distinguish between different
  shower types (as discussed below) and to improve the energy estimation of the primary particle as well as the angular resolution. Thus, IACTs record the Cherenkov
  light emitted by the shower with a fine-pixelated camera of 500–2000 PMTs. The

field of view (FOV) of the individual pixels is about 0.1°, while the overall FOV of the camera is typically ~4°.

- Multiple images of the same air shower seen from different angles also help to distinguish gamma-ray induced showers from hadronic showers, and provide a better energy and angular resolution. IACT observatories consist therefore of at least 2 telescopes.
- A fast integrating electronics, which is able to sample the signal at a rate larger than 5×10^8 Hz, keeps the contribution of background light in the images low. This is especially important when trying to achieve a low energy threshold.

A sketch of the general IACT technique is given in Fig. 3.6. Another important in-gredient to efficiently obtain images of air showers, is the trigger system. Normally, the decision to record the sampled signal of the PMTs follows a three-level concept: (i) a single pixel must show a minimal fluctuation to pass the first level; (ii) either a group of neighboring pixels or a given number of pixels in a pre-defined area must pass the first level; (iii) between a given time window at least two telescopes must report a level two trigger. This trigger system makes sure that light flashes produced by fluctuations in the night sky background (NSB), as well as from scattered light, are discarded, since they do not display the tight spatial and temporal correlations of Cherenkov flashes associated with EAS. This restriction is especially crucial con-sidering that these Cherenkov flashes contribute merely ~0.1% to the total light of the night sky [13]. Thus, with the goal of detecting gamma rays from the ground, the overwhelming background for IACTs at trigger level are air showers initiated by cos-mic rays. Even for the strongest steady source of VHE gamma rays, the Crab nebula, the number of recorded hadronic showers usually exceeds the number of gamma-ray showers by a factor of ~10^4.

In the 1980s, Hillas [15] proposed the pioneering concept to distinguish gamma-ray EAS from hadronic showers with IACTs. With the help of Travor Weekes, he suggested that the resolved image of an EAS could be modeled by a two-dimensional ellipse as shown in Fig. 3.6. Through Monte Carlo (MC) simulations he showed that with the help of the corresponding parameters, which today are known as the *Hillas parameters*, one is able to statistically discriminate between both types of showers. Later on these parameters were complemented by stereoscopic and timing parameters, whose discrimination power was demonstrated by the HEGRA experiment [16] and the MAGIC telescopes [17], respectively. In the following we list some important parameters and provide a short description (see also Figs. 3.6 and 3.7):

- *Size*: Total amount of light in the shower image, measured in photo electrons (ph.e.).
- *Length* and *Width*: The RMS spread of the photo electrons along the major and minor axis of the ellipse.
- *Distance (Dist)*: Angular distance between the assumed source position and the center of gravity (COG) of the image (taking the photo electrons as weights).
- *Alpha*: The angle between the major axis of the ellipse and the *Dist* parameter.

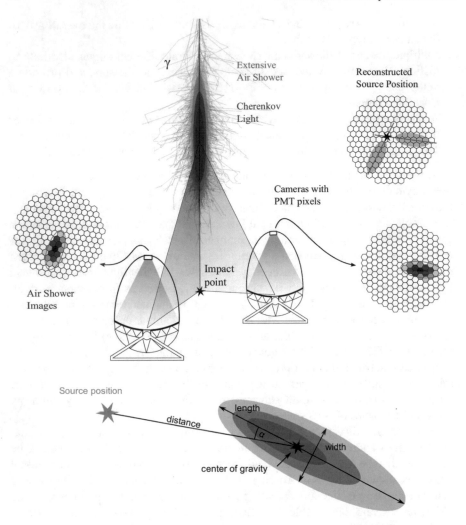

Fig. 3.6 Sketch of the general IACT technique. **Top**: Various telescopes record the Cherenkov light of the air shower by means of PMTs. The shower images of the telescopes can be combined to obtain powerful stereo parameters (see text and Fig. 3.7). From the impact point and the reconstructed origin of the gamma ray in the sky (also called reconstructed source position), one is able to reconstruct the shower axis. **Bottom**: The shower images induced by cosmic gamma rays tend to resemble ellipses and are usually parametrized by the so-called *Hillas* parameters (see text). Figures taken from [14] (Color figure online)

- *Time gradient*: The coefficient of the linear term of the fit to the arrival times along the major axis. As shown previously in Fig. 3.5 (right panel), its sign does not only depend on the direction of the source position but also on the *Impact* parameter.
- *Impact*: Shortest distance between the shower axis and the telescope.

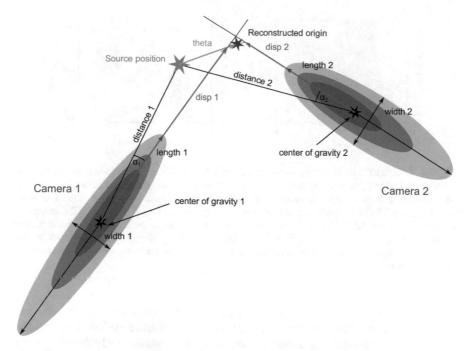

Fig. 3.7 Illustration of some *Hillas* and stereo parameters. The shower origin (also called reconstructed source position) is marked by a blue star and can be reconstructed by the crossing point of the major axes of the shower images, or by means of the *Disp* parameters (see Sect. 3.4). The assumed source position is marked by an orange star. Figure taken from [14] (Color figure online)

- *Maximum Height* (*MaxHeight*, h_{max}): Height of the brightest part of the shower above the ground.
- *LeakageN*: The fraction of the *Size* parameter contained in the N outermost rings of pixels in the camera. A high *Leakage* means that the shower image is substantially truncated and the estimation of the image parameters becomes unreliable.
- *Disp*: Distance from the COG to a reconstructed source position that lies on the major axis of the corresponding ellipse. Ideally this reconstructed source position coincides with the crossing point of the major axes from various shower images. Section 3.4 will give more details on the calculation of this parameter.
- *Theta* (Θ): Angular distance between the assumed source position and the final reconstructed source position.

Two out of those image parameters are especially helpful when trying to distinguish between gamma-ray and hadronic showers: the *Width* and the *Maximum Height* (see Fig. 3.8). As discussed in the previous section, hadronic showers exhibit a diffuser structure and yield on average broader and more elongated images than gamma-ray showers. Hence, *Width* (and also the *Length*) are good discrimination parameters especially for large image sizes (higher energies). At lower energies small muon events and electromagnetic subcascades in the hadronic showers can easily

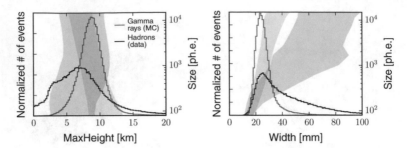

Fig. 3.8 Distribution of the image parameters *MaxHeight* and *Width* as measured by the MAGIC telescopes. The left *y*-axes correspond to the step histograms, the right *y*-axes to the shaded areas. The areas denote the mean of the distribution plus-minus its variance in function of the *Size* of the shower image. The histograms represent the distributions for *Sizes* >200 ph.e. **Left**: The parameter *MaxHeight* is especially effective for smaller images. An additional peak appears at lower heights for image sizes $\lesssim$700 ph.e. as a result of single muon events triggering multiple telescopes. **Right**: The *Width* parameter reflects the diffuse and broad structure of hadronic showers at higher energies (Color figure online)

mimic gamma-ray events and become the dominant residual background. In addition, the low *Size* of low energy events make the estimation of the image parameters less precise and further aggravate the gamma/hadron separation. On the other hand, the separation power of the *Maximum Height* parameter seem to increase with lower image sizes. For gamma-ray showers the *MaxHeight* distribution resembles a single Gaussian, which shifts to lower heights for increasing energies since high-energy showers are able to penetrate deeper into the atmosphere. For hadronic events, the distribution is broader and gets more complicated at lower sizes. An additional peak appears at lower heights, which is a product of single muon events triggering various telescopes.

For a more detailed description of the IACT technique and its historical development the reader is referred to the lecture notes of Weekes [18] and the extensive review by Lorenz and Wagner [19].

3.3   The MAGIC telescopes

After the dismantling of the HEGRA experiment in 2002, scientists of the Max-Planck Institute (MPI) for physics in Munich (Germany) were joined by Spanish and Italian research groups to build the Major Atmospheric Gamma-ray Imaging Cherenkov (MAGIC) telescopes in La Palma, Canary Islands (Spain). The chosen site was the Roque de los Muchachos Observatory (ORM) situated at about 2200 m above sea level. Unlike the H.E.S.S. and the VERITAS collaboration, the MAGIC collaboration included only one telescope in their original design but made it the largest back then in terms of the reflector size. The 17 m in diameter mirror dish was necessary to achieve a lower energy threshold of below 50 GeV and to close the gap between space- and

Fig. 3.9 The two MAGIC telescopes at the Roque de los Muchachos Observatory in La Palma, Canary Islands (Spain). Picture credits: Robert Wagner. Picture taken from http://magic.mppmu. mpg.de/gallery/pictures/, last accessed 04/10/2017

ground-based gamma-ray observations, a primary goal for the MAGIC collaboration. The second goal was the ability of a fast repositioning, and hence a light weight design, to catch transient events, such as Gamma-Ray Bursts (GRBs). The MAGIC telescope was inaugurated in 2003 and was joined by a second one, basically a clone of the first one, in 2008 (Fig. 3.9). We will first describe the components of the current system as of 2018 and then talk about the major upgrades of the instrument that have been performed since the start of its operation and that will become important for the data analysis in Chap. 5 later on.

3.3.1 Drive System

The telescope has an alt-azimuth mount, tracking a source in the sky by moving around two axes (azimuth and elevation). The azimuth axis is equipped with two 11 kW motors, while the elevation axis has a single motor of the same power. The telescope position can be determined with an accuracy of ~0.02° with the help of three shaft encoders. The optical axis of the telescope is then calibrated by means of two CCD cameras: (i) the T-Point camera takes pictures of about 150 different stars (different azimuth and zenith angles) in order to produce a *drive bending model*, which takes into account distortions of the telescope frame at different pointing positions; (ii) the starguider camera constantly monitors the pointing position to take into account irregular bendings of the telescope structure or changes in the bending with time. Together, the starguider and the shaft encoders, can achieve a maximum pointing accuracy of 0.01°.

3.3.2   Reflector

The telescope mirror support frame is built of robust, light weight carbon fiber tubes. The frame plus the mirrors and the camera support structure weighs less than 20 tons enabling a fast movement of the telescope. The reflector follows a parabolic profile with a diameter of 17 m. The parabolic shape preserves the timing between the Cherenkov photons coming from an air shower, which is important at the trigger level to reduce accidental triggers by the NSB. On the other hand, the parabolic shape has the downside of a coma aberration for off-axis showers. The reflector itself consists of 964 square mirrors of ~ 0.25 m^2 in the first telescope (MAGIC-I), grouped into support panels of 4 mirror each, and 247 mirrors of ~ 1 m^2 in MAGIC-II [20]. The mirrors yield approximately $\sim 85\%$ reflectivity and, depending on their location in the dish, have a slightly different curvature to match the parabolic profile. A dedicated active mirror control (AMC) hardware and software enables mirror adjustments in function of the zenith angle of the observation, which are necessary due to small distortions of the telescope frame. The optical point spread function (PSF) of the instrument can be checked by images of bright stars projected on a dedicated lid in the center of the camera. Theses pictures are taken with a camera from the Santa Barbara Instrument Group (SBIG, [21]). Another way to estimate the optical PSF is to record the Cherenkov light from muon events [22]. The muon ring images are normally triggered by only one telescope at a rate of about ~ 2 Hz, and hence are subject of special observations called *muon runs*. The comparison of the muon parameters with MC simulations yield important parameters needed for the correct simulation of the instrument response, in particular the overall light collection efficiency of the telescopes, which includes the mirror reflectivity.

3.3.3   Camera

The geometry of the camera is shown in Fig. 3.10. It has a roundish shape with a field of view (FOV) of $\sim 3.5°$ and is equipped with 1039 PMTs grouped into clusters of 7 to form a modular unit for easier installation and maintenance. To reduce the dead space between the round photocathodes of the PMTs, hex-to-round light guides, so-called Winston cones, are set on top of the PMTs, which also help to reduce the contamination of scattered light. The gains of the single pixels (PMTs) show a large spread from 1×10^4 to 6×10^4 measured at 850 V. Such a spread is not unusual for PMTs and is compensated by adjusting the applied high voltages (HV) individually with a *flatfielding* procedure (see [23] for details).

3.3.4   Data Acquisition

The signal transmission from the PMTs to the data acquisition PC (DAQ PC) is schematically shown in Fig. 3.11. The pre-amplified PMT signals are converted into light using

Fig. 3.10 Picture and geometry of the MAGIC cameras. **Left**: The picture shows the MAGIC-II camera with open lids. Picture credit: Robert Wagner. **Right**: The MAGIC cameras are made out of 1039 PMT pixels equipped with hexagonal Winston cones (see text). The clusters of 7 pixels are indicated by thick black lines. The cyan hexagons denote the L1 trigger macrocells, with overlapping pixels marked as green (single overlap) and red (double overlap). Figure taken from [23] (Color figure online)

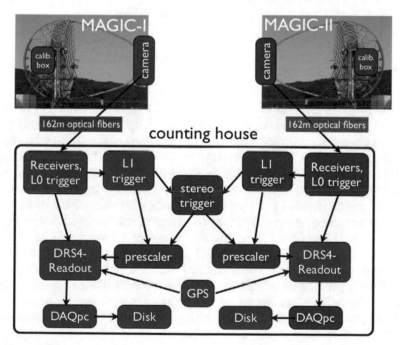

Fig. 3.11 Sketch of the data acquisition and the MAGIC trigger system (see text for details). Figure taken from [23]

fast current driver amplifiers coupled to vertical cavity surface emitting laser diodes (VCSELs). The optical signal is then transmitted from the camera through 162 m long optical fibers to the counting house. The conversion of the electrical signal back into an optical one has several advantages: (i) signal attenuation and dispersion are much lower for an optical fiber than for a coaxial cable; (ii) optical fibers weigh much less than coaxial cables; (iii) there is no electromagnetic pickup. In the counting house the optical signal is converted back into an analog electrical signal and split into two by the receiver boards and routed to the trigger branch and the signal digitalization branch. The receiver boards also generate the Level-0 (L0) trigger signal (see the following paragraph about the trigger system). The analog signal is digitized by the DRS4-readout based on the DRS4 chip that samples the signal at 1.64 GHz and writes it into a ring buffer built of 1024 switching capacitors ([24], and references therein). In the event of a trigger the buffer is read out by a conventional analog to digital converter (ADC). The final data acquisition to the storage disks is performed by the DAQ PCs via a multithread C++ program. The signal from the PMT to the final readout has to be calibrated, that is the measured ADC counts have to be converted into the physical quantity of photoelectrons as well as the ADC timing has to be connected to an absolute signal timing. For this reason calibration boxes are installed in the center of the two reflector frames, which are able to uniformly illuminate the whole camera with well-characterized light pulses of different intensities.

3.3.5   Trigger System

Only the 547 inner pixels of the camera are able to trigger the readout, as indicated in Fig. 3.10. The receiver boards check if the signal amplitude of one of these pixels exceeds a given discrimination threshold (DT) and forward the digital L0 signals to the telescope trigger L1. The trigger rate of individual pixels (Individual Pixel Rate, IPR) is dominated by the night sky background (NSB) and typically falls in the range of 300–600 kHz. Bright stars in the field of view can easily increase this rate by a factor of 2 and a individual pixel rate control (IPRC), which adjusts the DT for single pixels, has proven to ensure a flexibility for different NSB levels while keeping the energy threshold low [23]. The L1 trigger consists of 19 macrocells of 36 pixels each with an overlap of one pixel row (see Fig. 3.10). In each macrocell a compact 3 next-neighbor (3NN) logic of L0 triggers is checked and a L1 trigger signal is issued if any of the macrocells report a positive check of the logic. The L1 signal is sent to the stereo trigger L3, which checks for coinciding L1 triggers from both telescopes, and to the *prescaler*, which ultimately decides which type of trigger provokes the readout. Under standard observations the prescaler only forwards L3 triggers and triggers from pedestal and calibration events, which are necessary for the signal calibration (see the signal calibration part in Sect. 3.4).

3.3.6 Atmospheric Monitoring

Since weather phenomena can heavily affect the performance and accuracy of IACTs, it is important to monitor atmospheric conditions during data taking. The principal instrument for this purpose is a Light Detection And Ranging (LIDAR) system that is able to measure the transmission profile of the atmosphere. It consists of a light detector that records the backscattered light of a laser shooting in the direction of the observation (see [14], for details). Another available instrument to judge the atmospheric condition, is the *Pyrometer* that measures the intensity of infrared radiation in the FOV of the MAGIC telescopes. These measurements indicate the temperature of the sky. If a cloud or haze covers the sky, the thermal radiation from the Earth's surface is reflected and the sky temperature increases. While performing long term measurements, a minimum and maximum temperature were found allowing for a definition of the *cloudiness* of the sky, for which the value 100 corresponds to the maximum temperature and 0 to the minimum. A couple of smaller instruments also provide information by visualizing the night sky or measuring the temperature, humidity and wind speed. All this information is gathered by a program called the *weather station* and provides feedback to the operators of the telescopes during the data taking.

Most of the components and subsystems of the telescopes can be operated via a central control software called *SuperArehucas* (SA, for details see [21, 25]). It is written in LabView and automatizes most of the complex tasks during observations, while providing a flexible unified graphical user interface for the operators of the telescopes.

3.3.7 The Major Upgrades of Magic

Since the inauguration of the first telescope in 2003, the MAGIC collaboration issued a serious of major upgrades to steadily improve the performance of the instrument and to extend the lifetime of the experiment (see Fig. 3.12). The initial camera of MAGIC-I had a hexagonal shape with 577 pixels of two different sizes due to a compromise between cost and performance. The smaller inner 397 pixels were complemented by 180 bigger outer pixels, which did not contribute to the trigger area [26]. In the first years of operation, the PMT signals were digitized at a sampling rate of 300 MHz via a system of flash analog to digital converters (FADCs) developed at the University of Siegen [27]. In February 2007 the data acquisition was upgraded with a faster FADC based on the Fiber-Optic Multiplexing (MUX) technique and capable to digitize the signal at 2 GSamples/s [28]. With the increased timing resolution of the signal and the use of this timing information in the analysis, the sensitivity of the instrument could be enhanced by a factor of 1.4 [29].

The construction and commissioning of the second telescope, a close clone to MAGIC-I, finished in autumn 2009. The main differences between MAGIC-I and II at

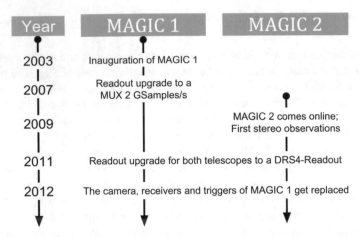

Fig. 3.12 Time line of the major upgrades of the MAGIC telescopes since 2003 (see text)

that time were the coarse- and fine-pixelated cameras (see above) and the readout, for which MAGIC-II used a 2 GSamples/s digitizer based on the DRS2 chip [30]. From 2011 to 2012 the telescopes underwent a major upgrade to make the stereoscopic system uniform, to ease its maintenance and to improve the overall performance. The readout of the two telescopes were upgraded to an improved DRS4-readout with less dead time, and in particular the camera of MAGIC-I, together with the receivers and the trigger, were replaced to match the hardware of MAGIC-II [23, 31]. In the end of 2014, the sampling frequency of the readout was lowered to 1.64 GHz, since for certain orientations of the telescopes the ring buffer of the readout was not large enough to wait for the L3 trigger. This occurred when the time of flight differences of the Cherenkov photons to the two telescopes were the largest, which created a L3 *dead zone* in the sky.

For a detailed description of the current hardware of the MAGIC telescopes the reader is referred to [23] and references therein. The hardware and performance of the pre-upgrade systems are discussed in [26, 29, 30]. In this thesis we will analyze data taken by the MAGIC telescopes from February 2007 on, that is after the readout upgrade of MAGIC-I to 2 GSamples/s.

3.4 MAGIC Analysis Chain

Data taken by the MAGIC telescopes is normally analyzed by means of the MAGIC analysis and reconstruction software (MARS, [32]). It is a software package written in C++ on top of ROOT [33], that provides the user with all procedures, algorithms and methods to complete the whole analysis chain, from reading MAGIC raw data to extracting high level analysis products, such as the spectrum of a gamma-ray source. MARS provides container classes to store data and classes that hold analysis algorithms, as well as many executables that run part of the analysis and that are called consecutively by the user via a command line. In general three input data streams are needed to obtain high level analysis products from MAGIC observations (see Fig. 3.13): (i) data, in which the gamma-ray signal is thought to be present; (ii) data, from which the background will be estimated; (iii) Monte Carlo (MC) simulations to estimate the response of the instrument.

The shower simulation and the simulation of the instrument response is done outside of MARS by a customized version of CORSIKA [34], called *Mmcs*, and two programs called *reflector* and *camera* [35], respectively.[3] Each new hardware configuration of the MAGIC telescopes requires a new set of MCs to reflect the updated response of the instrument. Not only human intervention may cause long-lasting alterations in the instrument response but also unfavorable weather. On the one hand, strong winds together with rain and snow can damage the AMC and worsen the optical PSF. On the other hand, Calima[4] with a high dust concentration, is able to reduce the reflectivity of the mirrors. Therefore, the optical PSF and the mirror reflectivity are monitored by means of 3DIG pictures and muon runs (see Sect. 3.3). Additionally, the overall performance of the system is checked regularly by measuring the Crab nebula flux, which serves as a standard candle, during the Winter season. Normally the MCs are generated at the TU Dortmund University and provided to the MAGIC collaboration through the Port d'Informació Científica (PIC) in Barcelona, which also hosts the MAGIC data center.

Depending on the observation mode of the telescopes, the signal and the background can actually be extracted from the same data. In the so-called *wobble mode* (or False Source Tracking method, [37]), two or four opposite directions with an offset $d \geq 0.4^\circ$ from the source, are tracked alternately for 20 min or less each, as illustrated by Fig. 3.14. The advantage of this mode is that the sky regions lying on the circle with radius d have the same exposure as the source region and can be used for the background estimation, since no gamma-ray events are expected from these regions. A common choice is to use the *anti-source* position, located farthest away, at 180°, from the source position in the camera plane, but further regions can be included if one makes sure that they are not contaminated by the gamma-rays from the signal region. The alternate tracking, or *wobbling*, between 2 and 4 positions, assures that possible radial asymmetries in the camera acceptance cancel out. Most MAGIC data is taken in this observation mode because, firstly no extra time is needed for background observations and secondly the background is measured simultaneously under the same weather and NSB conditions. The draw back is a slightly reduced trigger acceptance due to the offset of the gamma-ray source from the center of the camera.

The other observation mode is the On-Off mode, in which the telescopes point directly to the source, so that the source position falls right into the camera center. In contrast to the wobble mode, in this mode additional Off observations are necessary to

---

[3] A new program called *matelsim* has been developed to replace *reflector* and *camera* [36].

[4] These are dust winds originating in the Saharan Air Layer especially during summer [14].

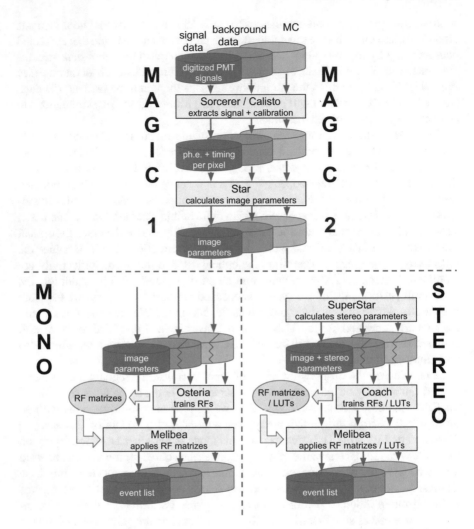

Fig. 3.13 Workflow of the low- and intermediate-level analysis chain of MAGIC. For both, mono and stereo analysis, the steps of the low-level analysis are the same. The low-level analysis, as well as the calculation of the stereoscopic parameters in the case of a stereo analysis, is usually done by the MAGIC on-site analysis (OSA) system installed on a computer cluster in the counting house (see Appendix C for details). The analysis products of OSA are then transferred to the MAGIC data center at the Port d'Informació Científica (PIC) in Barcelona. The user typically starts the analysis of MAGIC data with downloading SuperStar data and the data quality selection (see text)

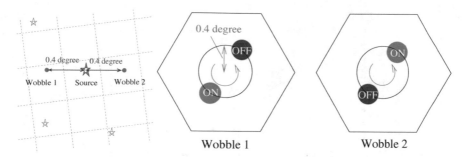

Fig. 3.14 Schematic explanation of the wobble observation mode. In order to compensate non-radial camera inhomogeneities, the wobble position is switched periodically. In the standard observation mode, MAGIC switches between 4 wobble positions, in which position 3 and 4 rotate the ON region by +90° and −90°, respectively, along the 0.4° circle. Figures taken from [38]

estimate the background. Ideally these observations should be conducted in a nearby sky region with a similar NSB level and temporally close to the On observations. The upside is that the On mode maximizes the acceptance and effective area for the source position being in the center of the camera, and therefore is advantageous if the background can be estimated from the the same sky region as the signal, such as in pulsar observations (see the subsection *Signal detection* below). An On-Off observation mode might also be necessary if the source extension is too big to allow for observations in wobble mode.

3.4.1 Signal Extraction and Calibration

The first step in the MAGIC analysis chain is the extraction and calibration of the digitized PMT signals (see Fig. 3.13). For this task two executables are available in MARS: Sorcerer, which is able to deal with DRS2 and DRS4 data, and a precursor Calisto, which can be used for older readout data (MUX and Siegen). To extract the light pulses of the PMTs (see Fig. 3.15), the baseline of the readout is obtained by means of *pedestal events* that are triggered at a fixed frequency, and should therefore only contain noise. From the integrated light pulse, two basic quantities have to be calibrated: (i) FADC counts have to be converted into photoelectrons (ph.e.); (ii) FADC slices have to be connected to an absolute signal timing. The conversion factor of FADC counts to ph.e. is calculated via the F-Factor method (also called Excess-Noise method, see [39, 40] for details), which is applied to calibration events generated by the calibration box. Calibration events also allow to connect the FADC slices to an absolute arrival time of the signal. Dedicated pedestal and calibration runs are taken before starting the observation of a new source. To counteract a changing NSB during observation and electronic instabilities (for example from the VCSELs), the baselines and conversion factors are updated every ∼30 s by means of interleaved pedestal and calibration events. These interleaved events are triggered at a frequency of 25 Hz each.

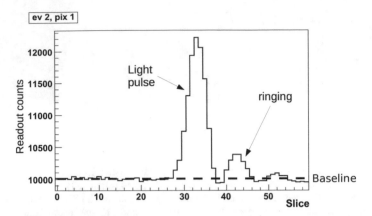

Fig. 3.15 A typical PMT signal digitized by the readout. The readout counts are plotted versus the readout slices, which correspond to the switching capacitors. The so-called *ringing* after the light pulse is an artifact of the readout. Besides light pulses, the readout can also be triggered by so-called *afterpulses* from the PMTs, which are large amplitude signals caused by an ion accelerated back to the photocathode of the PMT. Image courtesy of Julian Sitarek

3.4.2   Image Cleaning and Parameter Calculation

After the extraction and calibration of the PMT signals we obtain an image of the shower that we want to parametrize as discussed in Sect. 3.2. The shower image does not only contain the signal from the shower but also noise from the NSB, which hinders the accurate parametrization. Hence, only pixels are selected that most probably contain useful information about the shower. This step is often referred to as *image cleaning* and is illustrated in Fig. 3.16. Both, the image cleaning and the parametrization, are done by the executable Star. Here we will describe two image cleaning algorithms that are the most common algorithms utilized in the MAGIC analysis: the absolute cleaning [29] and the sum cleaning [41]. The basic principle of both algorithms is the fact that Cherenkov photons from air showers are mostly clustered on the camera plane, both spatially and temporally, while NSB photons are randomly distributed. Both algorithms use a two-step approach, in which the first step consists of finding core pixels and the second step checks for boundary pixels. In the absolute cleaning a pixel is defined as core pixel if its charge is above a certain threshold and its arrival time is within 4.5 ns of the mean arrival time of all core pixels weighted by their charge. In the sum cleaning the core pixels are determined by summing up the clipped charge[5] in groups of next neighbors (2NN, 3NN and 4NN) and by comparing the sum to a given threshold $Q_{x\text{nn}}$ (with $x = 2, 3, 4$). Additionally the arrival time of the pixel has to lie within a time window $\Delta t_{x\text{nn}}$ from the mean arrival time of the xNN group. The additional parameters of the sum cleaning were

---

[5]The clipping prevents a dominating contribution from one pixel to the sum caused by, for example, PMT afterpulses.

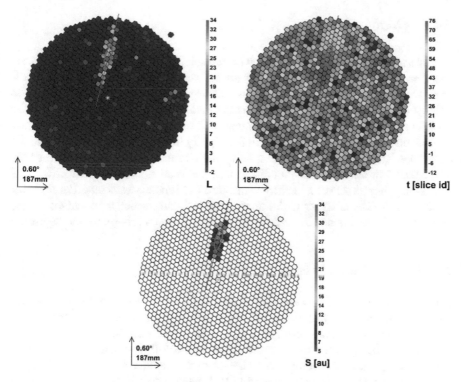

Fig. 3.16 The cleaning procedure applied to a typical shower image recorded by the MAGIC-II camera. **Left**: The raw shower image in which the color scale denotes the charge of each pixel. **Right**: The color scale shows the arrival time (in units of the readout slice) of the light pulse in each pixel. The light pulses from the shower exhibit a temporal clustering. **Bottom**: With the charge and time information we are able to clean the image (see text), fit the surviving pixels with an ellipse (green line) and extract the image parameters. The red line depicts the major axes of the shower ellipse. Figures taken from [21] (Color figure online)

extensively optimized and their numerical values are summarized in [25]. Once the core pixels are found, both the absolute and the sum cleaning check for boundary pixels in the same way. If the charge of a neighboring core pixel is above a certain threshold and its arrival time is within 1.5 ns of the core pixel, then it is accepted as boundary pixel and also survives the cleaning. The sum cleaning shows a better performance especially at low energies, and is the standard algorithm in the MAGIC analysis chain [21, 42]. However, since its implementation occurred only in 2011, older data was still cleaned with the absolute cleaning algorithm.

After the cleaning, the core and boundary pixels are finally used to obtain the shower parameters as discussed in Sect. 3.2. In the case of stereo observations, the executable SuperStar combines the shower images from both telescopes and calculates the stereo parameters. The calculation of the shower parameters constitutes the last step of the MAGIC low-level analysis, which is usually carried out by the On-Site Analysis chain (OSA, see Appendix C).

3.4.3 Quality Selection

The intermediate-level analysis starts with the selection of good quality data by the analyzer. MAGIC data can be affected by advert weather conditions or hardware failures, which can lead to erroneous flux estimations and a worse sensitivity. Hardware failures are usually detected by a dedicated data check system that runs over the subsystem reports and the data itself after each observation night. Incidences and unusual behavior is also noted down in the nightly runbooks by the operators of the telescopes. The quality of the data regarding the weather can be judged with the help of the LIDAR information, the cloudiness and the weather station (see Sect. 3.3), as well as the shower rate or a healthy distribution of image parameters. For this task MARS provides the analyzer with the executable Quate, which summarizes all the necessary information and discards affected data based on criteria set by the user.

3.4.3.1 Event Reconstruction

For the high-level analysis, we need to reconstruct three main parameters of an event: the energy of the primary gamma ray, the direction of the shower and the *Hadronness* parameter. The *Hadronness* parameter is a measure of likelihood that the shower was initiated by a hadron and takes a value between 0 (less likely) and 1 (more likely). This parameter is computed by means of the Random Forest (RF) technique, a powerful event classification method [43]. It has become the standard tool in the MAGIC analysis to separate gamma-ray induced showers from the overwhelming background of hadronic showers. The RF technique is a supervised machine learning algorithm, and hence requires a training before its application. The training consists of "growing" a large number of multidimensional decision trees by feeding it a sample of events of known nature. Usually the training samples are a set of suitable MCs and a data sample of pure background. Each decision tree uses a set of shower parameters specified by the user to try to efficiently separate gammas and hadrons see [44], for details). By default MAGIC uses 12 parameters for the decision trees, of which the *Width* and the *MaxHeight* yield the largest separation power (see Fig. 3.8) The trained decision trees are stored as matrices that are applied to the data to obtain the *Hadronness* parameter for each event. It is important to note that the events used for the training must not be used in the further analysis, since the decision trees are naturally biased towards these events.

The same technique is applied to estimate the energy of the primary gamma ray in the case of a mono analysis (see Fig. 3.13). Since the energy of the primary particle is proportional in first order to the total Cherenkov light yield from a gamma-ray shower, the *Size* parameter is naturally decisive for the energy estimation. In the case of a stereo analysis, the energy estimation is done independently for each telescope by means of look-up tables (LUTs). The LUTs are binned in *Size* and *Impact* and filled with the known energy (also called true energy) of the corresponding MC events. The values in the LUTs are the mean and RMS of the E_{true} distribution in each bin. In the

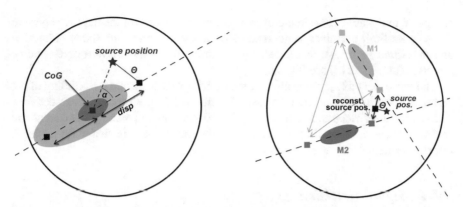

Fig. 3.17 Sketch of the *Disp* parameter and reconstruction of the origin (reconstructed source position). **Left**: For a perfectly reconstructed shower image, the origin of the shower should lie on the major axes of the ellipse. The absolute value of the *Disp* parameter is computed by means of the RF technique from MCs (see text). The *Alpha* and *Theta* parameter (α and Θ) are both calculated with respect to an assumed source position (denoted as a red star). Both values should be small if the shower originated from the direction of the assumed source. **Right**: For stereo observations the degeneracy of solutions for the *Disp* parameter (grey double arrows) can be solved by choosing the *Disp* pair with the smallest angular separation (black double arrow). Instead of choosing the crosspoint of the two major axes, we reconstruct the source position by choosing the weighted average between the two *Disp* positions (see text) (Color figure online)

end the estimated energy E_{est} of an event is the average of the estimated energies from both telescopes obtained by the corresponding bin in the LUT and weighted by their RMS. Furthermore small empirical corrections are applied, such as regarding the zenith angle and the *Leakage*.

To reconstruct the shower direction or origin (also called reconstructed source position or origin) the stereo analysis employs the DISP RF method. In this method a RF is trained for each telescope to determine the *Disp* parameter by means of MCs [45]. The training only allows to determine the absolute value of the *Disp* parameter, not its direction along the main axis of the shower image. This degeneracy of solutions is usually solved by combining the two shower images from both telescopes in the camera plane and choosing the *Disp* pair with the smallest angular separation (see Fig. 3.17). If all four possible separations are above 0.05 deg² the event will be discarded from the analysis. The final reconstructed direction is then taken as the average of the single *Disp* positions weighted by the number of pixels used in each shower image. This technique has proven to yield an improved angular resolution compared to the simple geometrical crossing method, in which the origin is reconstructed by the crossing point of the two main axes of the shower images. In addition this method also provides an additional gamma/hadron discrimination, since the estimation of the *Disp* parameter is optimized on gamma-ray showers and will often result in conflicting *Disp* parameters for hadronic showers. In principle the *Disp* method can also be applied to mono data but yields a worse sensitivity compared to the use of the simpler *Alpha* parameter.[6] Thus, for the mono analysis in Chap. 5, we limit ourselves to the major axis as estimation of the shower direction and the *Alpha* parameter as indicator if a gamma ray comes from a given source direction ($|Alpha| \simeq 0°$, see Fig. 3.6).

The two executables responsible for the training of the RFs and the building of the LUTs, are Osteria and Coach in the case of a mono and stereo analysis, respectively. The obtained matrices and LUTs are applied to the data and MCs by the executable Melibea. For a more detailed description of the mono analysis chain, the reader is referred to the author's master thesis.[7]

3.4.4 Signal Detection

Once we have an event list containing the *Hadronness*, energy and reconstructed source position for each shower, we can advance to check for a gamma-ray signal in our data. First, we will perform a cut based on the *Hadronness* parameter to reduce the number of hadronic events in our sample. The MAGIC collaboration established some recommended standard cuts that try to maximize the sensitivity depending on the energy. For the current system these cuts are:

- Low energy ($\lesssim$200 GeV): Hadronness < 0.28 & Size > 60
- Full range ($\gtrsim$200 GeV): Hadronness < 0.16 & Size > 300
- High energy ($\gtrsim$1 TeV): Hadronness < 0.1 & Size > 400 & Energy > 1 TeV

The presence of a gamma-ray source will lead to a peaked distribution of the squared Θ parameter towards zero, while for the background the distribution should be roughly flat.[8] In the case of wobble observations the Θ^2 distribution of the background is obtained by computing the Θ parameter with respect to the anti-source position. To increase the background statistics one can also choose further background positions that lie on the circle of the wobble offset. With an upper Θ^2 cut, which can be optimized on MCs or known gamma-ray sources like the Crab nebula, one defines the signal region from which the number of On and Off events are taken (see Fig. 3.18, left panel). Most commonly the significance of the signal is calculated by means of Formula 17 in [46]:

$$S\,(N_{\mathrm{on}}, N_{\mathrm{off}}, \alpha) = \sqrt{2} \left(N_{\mathrm{on}} \ln \left[\frac{1+\alpha}{\alpha} \left(\frac{N_{\mathrm{on}}}{N_{\mathrm{on}} + N_{\mathrm{off}}} \right) \right] \right.$$
$$\left. + N_{\mathrm{off}} \ln \left[(1+\alpha) \left(\frac{N_{\mathrm{off}}}{N_{\mathrm{on}} + N_{\mathrm{off}}} \right) \right] \right)^{\frac{1}{2}}, \quad (3.1)$$

---

[6]Under the assumption that the gamma/hadron RF is trained with source dependent and source independent parameters, respectively. See [45] for details.

[7]http://www.openthesis.org/document/view/603609_0.pdf, last accessed 15/04/2018.

[8]The falling acceptance of the camera towards the edges can introduce a negative slope in the Θ^2 distribution.

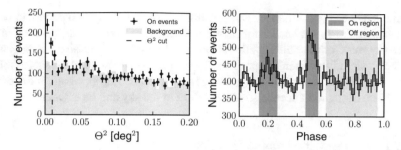

Fig. 3.18 Detections of a gamma-ray signal. **Left**: The plot shows the Θ^2 distributions for On and Background events in the case of a gamma-ray signal. The Θ^2 cut defines the signal region (see text). Following Eq. 3.1 the on counts deviate from the background model by more than 5 sigmas. **Right**: In the case of pulsar observations, the On and Off counts can also be extracted from the phaseogram. The phase regions for the On and Off counts should be defined a priori to avoid any selection bias (Color figure online)

where N_{on} and N_{off} are the On and Off counts, respectively, and α is a normalization factor that scales the number of Off events to match the On observation. In the case of On-Off observations, α would be the difference in observation time, in the case of wobble observations it is simply one divided by the number of used Off positions. In MARS the executable *Odie* takes care of producing Θ^2 plots and evaluating the significance.

For pulsar observations the signal can also be evaluated directly from the pulsar light curve. Since the signal is not only clustered spatially but also temporally, the background can be estimated from the same sky region as the signal but at different times corresponding to the Off phase region in the phaseogram (see right plot of Fig. 3.18). The signal counts are estimated from the On phase region, respectively. Both, the signal and the background region, should be defined a priori to avoid any selection bias.

3.4.5 Sky Maps

Sky maps display the binned event arrival directions in sky coordinates and are helpful to detect unexpected sources in the FOV of the observation or to study the morphology of extended sources. For IACTs the main challenge to produce sky maps is the correct estimation of the background. In MARS the background for sky maps is obtained via an exposure map model of the camera, that is the distribution of reconstructed shower directions in the camera plane. By default this model is constructed from the camera half corresponding to the anti-source position, but can also be gained by the "blind map" method, in which the lower value of the two camera halves is taken and corrected for the selection bias. For each event in the On map we sample 200 events from the exposure model to obtain a sufficiently precise background map. From the

On and the background map we can construct a test statistic (TS) sky map, in which we calculate the TS via Eq. 3.1 for each bin setting $\alpha = 200$. The distribution of TSs in the sky map for the null hypothesis closely resembles a Gaussian but in general can have a slightly different shape or width (see [42], for details and caveats of this TS map). In MARS the executable *Caspar* takes care of producing sky maps and provides some basic fitting tools for the source position in the sky and its extension.

3.4.5.1   Flux and Spectrum Calculation

To calculate the gamma-ray flux of a source we need to know the effective area of our detector. For IACTs the effective area (also referred to as collection area) is estimated via MC simulations and defined as

$$A_{\text{eff}} = A_{\text{sim}} \times \frac{N_{\text{sel}}}{N_{\text{sim}}}, \qquad (3.2)$$

where A_{sim} is the area above the telescope in which gamma-ray induced showers are simulated uniformly, N_{sim} is the total number of simulated gamma-ray showers and N_{sel} is the number of simulated gamma-ray events that trigger the telescope or survive the analysis chain (reconstruction and *Hadronness*/Θ^2 cuts), respectively. Hence, the ratio $N_{\text{sel}}/N_{\text{sim}}$ can be interpreted as shower detection efficiency of the instrument. The simulated area is usually a circle of roughly $r \gtrsim 350$ m depending on the zenith angle of the simulation (larger areas for higher zenith angles) and type of simulated particle (protons require a larger area than gamma rays). For zenith angles below $30°$ the effective area A_{eff} reaches values of $\sim 10^5$ m^2 depending on the energy as plotted in Fig. 3.19.

Apart from the number of excess events and the effective area, we also need to know the effective observation time t_{eff} to state a flux estimate. The effective time slightly differs from the observation time due to a dead time of the readout after each trigger. Assuming that triggers follow a Poisson distribution with a mean arrival time difference τ, the time differences dT between triggers follow an exponential function $A \exp(-dT/\tau)$. After fitting the dT distribution for τ, the effective observation time t_{eff} is obtained by multiplying the total number of events in the distribution with τ. In MARS the differential and integral flux, as well as a light curve, is calculated by the executable *Flute* [32].

The number of excess events, the effective area and the effective observation time allow us to calculate the spectrum of a source in estimated energy bins. Due to the finite energy resolution of the instrument, there may be large differences between the estimated and the true energy of an event. Thus, for a physical interpretation of the results, we need to correct for these differences, which is generally referred to as unfolding. The spectra provided by Flute already include a basic correction given an a priori assumed spectral shape of the source (casually referred to as "Poor Man's Unfolding", Zanin et al. [32]). Nevertheless, a full unfolding has to solve the equation

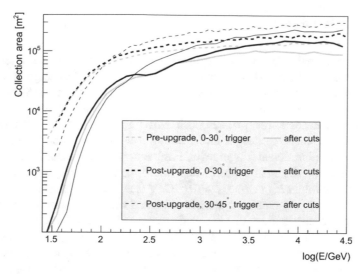

Fig. 3.19 Effective areas for two zenith ranges (0–30° and 30–45°) versus true energy at trigger level and after the whole analysis chain, that is after cuts. As reference we also show the effective area before the latest upgrade in 2011/2012 (see Fig. 3.12). Figure taken from [31]

$$Y_i = \sum_{j=1} M_{ij} \, S_j, \qquad (3.3)$$

where Y_i and S_j are the number of events in bin i of E_{est} and bin j of E_{true}, respectively. The migration Matrix M_{ij} represents the fraction of events in bin j of E_{true} moving into bin i of E_{est} due to the finite energy resolution, and is obtained via the MC simulations. Equation 3.3 represents a system of linear equations and its solution S can be obtained by minimizing[9] $\chi_0^2 = \sum_i \left(Y_i - \sum_j M_{ij} S_j \right)^2$. However, this minimization is not stable and leads to large fluctuations in S. The remedy to this problem is adding a regularization term to χ_0^2 that smoothly suppresses the fluctuations and permits a stable solution. In MARS the executable *CombUnfold* takes care of the full unfolding and implements several methods for the regularization, such as the Tikhonov, Bertero or Schmelling method, which are discussed in detail in Albert et al. [47]. Another way to solve Eq. 3.3, and also implemented in CombUnfold, is to constrain the distribution S by a parametrization and minimize χ_0^2 with respect to the free parameters of the a priori chosen parametrization. This is referred to as *forward folding* (or forward unfolding), which does not provide spectral points of the measurement but the best fit of the parametrization, such as a power law, with the corresponding errors. In general forward folding is a robuster method than unfolding with its need of a regularization mechanism. CombUnfold also allows to combine various outputs from Flute by simply adding the excess events and averaging over the instrument responses using the corresponding effective observations times as weights.

---

[9]Here we neglect the covariance matrix of Y, that is we set it to the identity matrix.

3.4.6   *Performance*

We will quickly revise the current performance of the MAGIC telescopes after the latest upgrades in 2011 and 2012 as stated in Aleksić et al. [31]. The energy threshold of a IACT system is commonly defined as the peak of the true energy E_{true} distribution of the MC gamma rays (see left panel of Fig. 3.20). Since low-energy showers are harder to reconstruct and also harder to distinguish between gamma-ray and hadronic showers, the energy threshold changes throughout the analysis steps being the smallest right after the calibration at trigger level. Furthermore, the threshold increases with zenith angle of the observation and its dependence can be roughly approximated by the empirical formula $E_{\text{th}} \propto 74 \cos(\theta)^{-2.3}$ GeV (see the right panel of Fig. 3.20).

The energy resolution of the system is estimated via the MC simulations. For this purpose the MC distributions of $(E_{\text{est}} - E_{\text{true}})/E_{\text{true}}$ in bins of true energy (5 bins per decade) are fitted with Gaussian functions and the resulting mean values and the standard deviations define the bias of the energy reconstruction and the energy resolution, respectively. Both quantities are plotted in the left panel of Fig. 3.21. For high energies the resolution degrades due to large, truncated images as well as worse statistics in the training sample for the LUTs. While for medium energies the resolution reaches its best value of ∼0.15, at lower energies it suffers from the challenge of precise image reconstruction. The energy bias increases rapidly at energies below ∼100 GeV due to the threshold effect, that is only upward fluctuations of low energy showers are triggered.

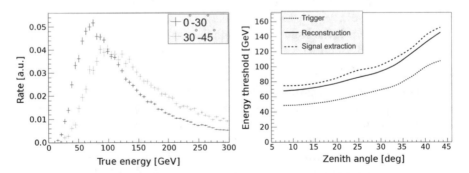

Fig. 3.20 Definition of the energy threshold and its dependence on the zenith angle. **Left**: The energy threshold is typically defined as the peak of the true energy distribution of MC events. The threshold increases with the zenith angle of the observation due to the increasing air mass between the shower and the telescopes. **Right**: The zenith dependence follows roughly the empirical formula $E_{\text{th}} \propto 74 \cos(\theta)^{-2.3}$ GeV. Low-energy showers are in general harder to reconstruct and are partially lost during the event reconstruction algorithm. They are also harder to distinguish from hadronic showers leading to a slight increase in the threshold after the application of all cuts (including *Hadronness* and Θ^2) and the extraction of the signal. Figures taken from [31]

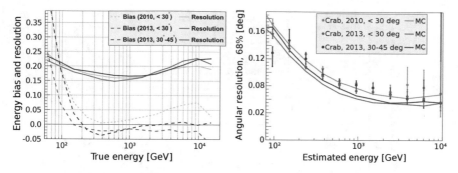

Fig. 3.21 Energy resolution/bias and angular resolution of the MAGIC telescopes before and after the latest upgrade, and for two ranges of zenith angles (0–30° and 30–45°). **Left**: Both, the energy resolution and bias, are energy dependent. The best resolution is achieved between a few hundred GeV and a few TeV. Both quantities (resolution and bias) are defined in the text. **Right**: The angular resolution of the MAGIC telescopes improves with energy and is roughly constant from a few TeV on. The different results obtained by Crab observations and by MC simulations, result in an additional 0.02° systematic error added in quadrature. Figures taken from [31] (Color figure online)

The angular resolution can be estimated via the MC simulations or via observations of a strong point source, such as the Crab nebula.[10] Figure 3.21 shows the 68% containment radius versus estimated energy obtained by the two methods. The angular resolution improves with increasing energy, since larger images are better reconstructed, and reaches a plateau of ∼0.04° above ∼2 TeV. The apparent differences of 10–15% between the two methods result in an additional 0.02° systematic error added in quadrature.

The evolution of the integral sensitivity of the MAGIC telescopes throughout the upgrades is illustrated in Fig. 3.22. It is computed from Crab nebula observations and is generally defined as the capability of providing a 5σ detection of a source in 50 h of effective observation time assuming that the source follows a Crab nebula-like spectral shape. In general the sensitivity at low energies (∼100 GeV) is limited by the background rejection power (meaning a low signal to noise ratio) and systematics in the background estimation. At the core energy range the limiting factor is the quality of the stereo reconstruction together with the effective area, while at energies above a few TeV the sensitivity boils down to the effective area covered by the telescopes and constraints in the stereo shower reconstruction [49]. Since the inauguration of the first telescope in 2003, the MAGIC collaboration has managed to improve the sensitivity of its instrument by an impressive factor of 5–10 depending on the energy range. The telescopes are currently able to detect a source at the flux level of ∼0.5% C.U. above ∼400 GeV in 50 h of observation time.

---

[10]It should be noted, however, that [48] recently measured a Gaussian extension of the Crab nebula of 52″ above 0.7 TeV with the H.E.S.S. telescopes.

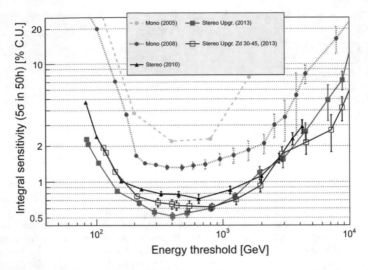

Fig. 3.22 Integral sensitivity of the MAGIC telescopes throughout the upgrades. The sensitivity is given in Crab units (C.U.) and is defined as $N_{\text{excess}}/\sqrt{N_{\text{background}}} = 5$ in 50 h of effective observation time, requiring $N_{\text{excess}} > 10$ and $N_{\text{excess}} > 0.05\,N_{\text{background}}$. Sensitivities are for observations below 30° in zenith, unless otherwise stated. For the corresponding upgrades, see Fig. 3.12. Figure taken from [31] (Color figure online)

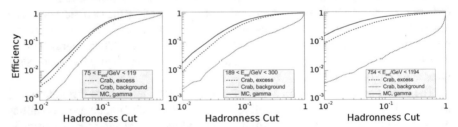

Fig. 3.23 Cut efficiencies for real data (Crab) and MC simulations for three different energy bins. The efficiency is defined as number of events surviving the *Hadronness* cut divided by the total number of events before the cut. For the highest energy bin the background rejection power is best as indicated by the separation of dashed and dotted lines. The discrepancies between data and MCs become visible for tighter *Hadronness* cuts (solid and dashed lines) and get more pronounced for higher energies. Figure taken from [31]

The systematic uncertainties for the IACT technique are an interplay between many small factors, some of them variable in time. Arguably the most important factors are the atmospheric transmission, conversion coefficient from photons to detectable photoelectrons and discrepancies between data and MC simulations. These discrepancies are illustrated in Fig. 3.23, in which the cut efficiency versus the *Hadronness* cut is plotted for three different energy ranges. Therefore, to keep this systematic uncertainty as low as possible, loose *Hadronness* cuts are usually chosen when reconstructing the spectrum of a source. For the current system the overall systematic uncertainties are estimated to be 11–18% for the flux normalization and around

15–17% in the energy scale [31]. A detailed list of all the single components, which contribute to the uncertainties, is given and discussed in Aleksić et al. [30].

For a detailed description of the instrument's performance at the corresponding time, the reader is referred to Albert et al. [26], Aliu et al. [29], Aleksić et al. [30, 31].

References

1. Weekes TC et al (1989) Observation of TeV gamma rays from the Crab nebula using the atmospheric Cerenkov imaging technique. Astrophys J 342:379. https://doi.org/10.1086/167599
2. Wakely SP, Horan D (2007) TeVCat: an online catalog for very high energy gamma-ray astronomy. In: 30th international cosmic ray conference, Merida, Mexico
3. Groom DE, Klein SR (2000) Passage of particles through matter. Eur Phys J C 15(1–4):163–173. https://doi.org/10.1007/BF02683419
4. de Naurois M, Mazin D (2015) Ground-based detectors in very-high-energy gamma-ray astronomy. CR Phys 16(6–7):610–627. https://doi.org/10.1016/j.crhy.2015.08.011
5. Atwood WB et al (2009) The large area telescope on the Fermi gamma-ray space telescope mission. Astrophys J 697(2):1071–1102. https://doi.org/10.1088/0004-637X/697/2/1071
6. Cherenkov PA (1934) Visible emission of clean liquids by action of gamma radiation. Doklady Akad Nauk SSSR 2(451)
7. Tamm IE, Frank IM (1937) Coherent radiation of fast electrons in a medium. Dokl Akad Nauk SSSR 14(107)
8. Mazin D (2007) A study of very high energy gamma-ray emission from AGNs and contraints on the extragalactic background light. Ph.D. thesis
9. Barrio JA et al (1998) The MAGIC telescope: design study for the construction of a 17 m Cerenkov telescope for Gamma-Astronomy above 10 GeV. Technical report, MPI
10. Wagner RM (2006) Measurement of very high energy gamma-ray emission from four blazars using the MAGIC telescope and a comparative blazar study. Ph.D. thesis, TUM
11. Jelley J, Galbraith W (1953) LXV. Light pulses from the night sky. Lond Edinb Dublin Philos Mag J Sci 44(353):619–622. https://doi.org/10.1080/14786440608521039
12. Anderhub H et al (2013) Design and operation of FACT—the first G-APD Cherenkov telescope. J Instrum 8(06):P06008–P06008. https://doi.org/10.1088/1748-0221/8/06/P06008
13. Blackett PMS (1948) A possible contribution to the night sky from the Cerenkov radiation emitted by cosmic rays. The Emission Spectra of the Night Sky and Aurorae, Papers read at an International Conference held in London, July, 1947. London: The Physical Society, p 34
14. Fruck C (2015) The Galactic Center resolved with MAGIC and a new technique for Atmospheric Calibration. Ph.D. thesis
15. Hillas AM (1985) Cherenkov light images of EAS produced by primary gamma rays and by nuclei. In: 19th international cosmic ray conference, La Jolla, USA, p 445
16. Kohnle A et al (1996) Stereoscopic imaging of air showers with the first two HEGRA Cherenkov telescopes. Astropart Phys 5(2):119–131. https://doi.org/10.1016/0927-6505(96)00011-4
17. Aleksić J et al (2010) Search for an extended VHE γ-ray emission from Mrk 421 and Mrk 501 with the MAGIC Telescope. Astron Astrophys 524:A77. https://doi.org/10.1051/0004-6361/201014747
18. Weekes TC (2005) The atmospheric Cherenkov imaging technique for very high energy gamma-ray astronomy
19. Lorenz E, Wagner R (2012) Very-high energy gamma-ray astronomy. Eur Phys J H 37(3):459–513. https://doi.org/10.1140/epjh/e2012-30016-x
20. Doro M et al (2008) The reflective surface of the MAGIC telescope. Nucl Instrum Methods Phys Res Sect A Accelerators Spectrometers Detectors Assoc Equip 595(1):200–203. https://doi.org/10.1016/j.nima.2008.07.073

21. Zanin R (2011) Observation of the Crab pulsar wind nebula and microquasar candidates with MAGIC. Ph.D. thesis, UAB
22. Goebel F et al (2005) Absolute energy scale calibration of the MAGIC telescope using muon images. In: 29th international cosmic ray conference, Pune, India
23. Aleksić J et al (2015) The major upgrade of the MAGIC telescopes, part I: the hardware improvements and the commissioning of the system. Astropart Phys 72:1–15. https://doi.org/10.1016/j.astropartphys.2015.04.004
24. Sitarek J et al (2013) Analysis techniques and performance of the Domino Ring Sampler version 4 based readout for the MAGIC telescopes. Nucl Instrum Methods Phys Res Sect A Accelerators Spectrometers Detectors Assoc Equip 723:109–120. https://doi.org/10.1016/j.nima.2013.05.014
25. Giavitto G (2013) Observing the VHE Gamma-Ray Sky with MAGIC Telescopes: the Blazar B3 2247+381 and the Crab Pulsar. Ph.D. thesis
26. Albert J et al (2008) VHE γ-Ray observation of the Crab Nebula and its Pulsar with the MAGIC Telescope. Astrophys J 674(2):1037–1055. https://doi.org/10.1086/525270
27. Stiehler R (2001) Konzeption, Entwicklung und Aufbau einer FADC-basierten Ausleseelektronik für das MAGIC-Teleskop. Ph.D. thesis, University of Siegen
28. Goebel F et al (2007) Upgrade of the MAGIC Telescope with a Multiplexed Fiber-Optic 2 GSamples/s FADC Data Acquisition system. In: 30th international cosmic ray conference, Merida, Mexico
29. Aliu E et al (2009) Improving the performance of the single-dish Cherenkov telescope MAGIC through the use of signal timing. Astropart Phys 30(6):293–305. https://doi.org/10.1016/j.astropartphys.2008.10.003
30. Aleksić J et al (2012) Performance of the MAGIC stereo system obtained with Crab Nebula data. Astropart Phys 35(7):435–448. https://doi.org/10.1016/j.astropartphys.2011.11.007
31. Aleksić J et al (2016) The major upgrade of the MAGIC telescopes, part II: a performance study using observations of the Crab Nebula. Astropart Phys 72:76–94. https://doi.org/10.1016/j.astropartphys.2015.02.005
32. Zanin R et al (2013) MARS, the MAGIC analysis and reconstruction software. In: 33rd international cosmic ray conference, Rio de Janeiro, Brazil, p 773
33. Brun R, Rademakers F (1997) ROOT—an object oriented data analysis framework. Nucl Instrum Methods Phys Res Sect A Accelerators Spectrometers Detectors Assoc Equip 389(1–2):81–86. https://doi.org/10.1016/S0168-9002(97)00048-X
34. Heck D et al (1998) CORSIKA: a Monte Carlo code to simulate extensive air showers
35. Majumdar P et al (2005) Monte Carlo simulation for the MAGIC telescope. In: 29th international cosmic ray conference, Pune, India, pp 203–206
36. López M (2013) Simulations of the MAGIC telescopes with matelsim. In: 33rd international cosmic ray conference, Rio de Janeiro, Brazil, p 692
37. Fomin V et al (1994) New methods of atmospheric Cherenkov imaging for gamma-ray astronomy. I. The false source method. Astropart Phys 2(2):137–150. https://doi.org/10.1016/0927-6505(94)90036-1
38. Saito T (2010) Study of the High Energy Gamma-ray Emission from the Crab Pulsar with the MAGIC telescope and Fermi-LAT. Ph.D. thesis, LMU
39. Mirzoyan R, Lorenz E (1997) On the calibration accuracy of light sensors in atmospheric Cherenkov fluorescence and neutrino experiments. In: 25th international cosmic ray conference, Durban, South Africa, vol 7, p 265
40. Gaug M (2006) Calibration of the MAGIC telescope and observation of gamma ray bursts. Ph.D. thesis
41. Rissi MT (2009) Detection of pulsed very high energy gamma-rays from the Crab Pulsar with the MAGIC telescope using an analog sum trigger. Ph.D. thesis
42. Lombardi S et al (2011) Advanced stereoscopic gamma-ray shower analysis with the MAGIC telescopes. In: 32nd international cosmic ray conference, Beijing, China
43. Breiman L (2001) Random forests. Mach Learn 45(1):5–32. https://doi.org/10.1023/A:1010933404324

44. Bock R et al (2004) Methods for multidimensional event classification: a case study using images from a Cherenkov gamma-ray telescope. Nucl Instrum Methods Phys Res Sect A Accelerators Spectrometers Detectors Assoc Equip 516(2–3):511–528. https://doi.org/10.1016/j.nima.2003.08.157
45. Sitarek J (2010) Gamma-rays from cascades in active galactic nuclei: observations with the MAGIC telescope and theoretical interpretation. Ph.D. thesis, University of Lodz
46. Li TP, Ma YQ (1983) Analysis methods for results in gamma-ray astronomy. Astrophys J 272:317–324
47. Albert J et al (2007) Unfolding of differential energy spectra in the MAGIC experiment. Nucl Instrum Methods Phys Res Sect A Accelerators Spectrometers Detectors Assoc Equip 583(2–3):494–506. https://doi.org/10.1016/j.nima.2007.09.048
48. Holler M et al (2017) First measurement of the extension of the Crab nebula at TeV energies. In: 35th international cosmic ray conference, Busan, South Korea
49. Hassan Collado T (2015) Sensitivity studies for the Cherenkov Telescope Array. Ph.D. thesis, UCM

Part II
Search for TeV Emission from the Crab and Other Pulsars

Chapter 4
MAGIC and the Crab Pulsar: History and Motivation

> *In this configuration, we detected pulsed γ-rays from the Crab pulsar that were greater than 25 giga–electron volts, revealing a relatively high cutoff energy in the phase averaged spectrum.*
>
> The MAGIC Collaboration, 2008

Since the commissioning of the first MAGIC telescope in 2004 (see Sect. 3.3), the MAGIC collaboration has always considered the Crab nebula and pulsar as valuable targets for observational campaigns and scientific research. To date approximately 14% of all the stored MAGIC data were taken on the Crab (nebula and pulsar). However, a significant portion of this 14% was taken as technical data to evaluate the performance of the instrument as described in Sect. 3.4, and had no immediate scientific purpose. A short summary of the scientific observational campaigns on the Crab is given in Chap. 5.

This chapter starts with a short introduction to the Crab nebula focusing on its central engine, the Crab pulsar. We then review previous observations of the Crab pulsar at very high energies (VHE) and emphasize MAGIC's role in it. We conclude the chapter with an answer to why a reanalysis and combination of all Crab pulsar data taken by MAGIC up to date, could provide new scientific results.

4.1 An Introduction to the Crab Nebula and Pulsar

From historical recordings it is safe to say that at least five supernovae were observed during the last millennium in our own galaxy [1]. One of them, supernova SN 1054, was first noticed by astronomy aficionados in the Far East and the Arabic world on the fourth of July in 1054 AD. On this day, Chinese chronicles speak of the appearance of a "guest star" near ζ Tauri, so bright that it was visible even in daylight for 23 days and only vanished from the night sky 21 months later. In 1942, after a thoroughly

© Springer Nature Switzerland AG 2019
D. Carreto Fidalgo, *Revealing the Most Energetic Light from Pulsars and Their Nebulae*, Springer Theses, https://doi.org/10.1007/978-3-030-24194-0_4

study of all ancient recordings, Mayall and Oort could confidently identify the Crab nebula and its pulsar with the supernova SN 1054, which also has been probably one of the brightest supernova on record so far [2].

The Crab (nebula and pulsar) is located at a distance of $\sim$2 kpc from Earth away from the Galactic Center [3] and is typically one of the brightest steady sources in the sky at photon energies above $\sim$30 keV. It is therefore a prime target to study non-thermal processes in our universe like polarized synchrotron radiation or pulsed emission from pulsars [4]. The broadband spectrum of the Crab nebula is composed of two broad non-thermal components (see Fig. 4.2 and Sect. 2.6). The first component is due to synchrotron radiation extending from radio to gamma-ray frequencies. While in radio the nebula extends out to $5'$ (3 pc) from the central pulsar the X-ray emission is confined within a bright torus of radius $40''$ (0.4 pc) surrounding the pulsar as can be seen in Fig. 4.1 [5]. The second emission component sets in above $\sim$400 MeV and is thought to be dominated by inverse Compton scattering, primarily of the synchrotron photons from the nebula itself [5, 6]. Due to the poor

Fig. 4.1 A composite image of the Crab nebula and pulsar observed in various energy bands. The picture combines data from 5 different telescopes: radio data from the Very Large Array (red), infrared data from the Spitzer Space Telescope (yellow), optical data from the Hubble Space Telescope (green), ultraviolet data from XMM-Newton (blue) and X-ray data from the Chandra X-ray Observatory (purple). The shape of the nebula is approximately ellipsoidal with a major axis of $\sim 7'$ and a minor axis of $\sim 4.6'$. Right in the middle of the nebula one can see the pulsar as a bright white dot. Image taken from and credits found at: http://hubblesite.org/image/4027/gallery, last accessed 22/09/2017 (Color figure online)

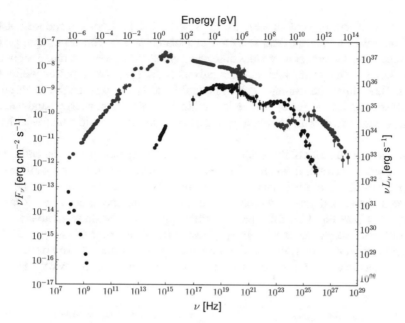

Fig. 4.2 The broadband spectral energy distribution of the Crab nebula (blue) and the Crab pulsar (black). For the nebula only the average flux is shown (excluding the flares, see text), for the pulsar the flux was averaged over the whole rotation phase. Figure adopted from Bühler and Blandford [4] (Color figure online)

angular resolutions of current generation gamma-ray telescopes of about $0.1°$ to $1°$, the emission region of greater than ~ 1 MeV photons has not been resolved yet but seems to encompass the pulsar and the torus/jet structure seen in X-rays [7, 8].

From hard X to gamma rays the Crab nebula is the only source in the sky that is bright and steady enough to be easily used as a normalization standard. Although recently it became clear that this might not necessarily be the case. From 2008 to 2010 several instruments detected a decline of $\sim 7\%$ in the overall Crab nebula flux in the 15–50 keV band and similar drops were found in the 3–15 keV and 50–100 keV data [9]. The biggest surprise, however, was observed in the high energy range where *Fermi* and AGILE have detected several gamma-ray flares from the nebula since 2008, a phenomena which has not been observed in any other pulsar wind nebula to date [5, 10]. During the brightest flare, lasting for approximately nine days in April of 2011, the gamma-ray flux doubled within eight hours and reached a peak photon flux 30 times higher compared to the average value [11]. At very high energies significant variability has not been detected yet allowing IACTs to use the Crab nebula as a *standard candle* to partially calibrate their instrument (see Sect. 3.4).

Not only the nebula showed recent surprises but also the Crab pulsar astonished scientists in the last years as will be discussed in detail in the next section. The pulsar was first discovered in 1968 at the Green Bank Observatory in West Virginia as a pulsating radio source by Staelin and Reifenstein [12]. In contrary to the first pulsar ever discovered earlier that year by Hewish et al. [13], Staelin and Reifenstein [12] reported an unusual time variation of isolated pulses with no unique period. Nowadays we know that they were observing *giant radio pulses*, a rare phenomena seen only in a few young and recycled pulsars so far [14]. These pulses with a flux up to 1000 times the average occur randomly and in the case of the Crab pulsar lead also to an increase of optical emission [15]. Links to other energy bands such as X-rays or gamma rays are under study, but a firm connection could not be established yet.

Another peculiarity of the Crab pulsar is its visibility throughout the entire electromagnetic spectrum, from radio frequencies up to very-high-energy gamma rays. Its pulse profile is characterized by two strong emission peaks: the main pulse P1, which at 1.4 GHz defines the rotation phase $\phi = 0$, and the inter-pulse P2 located at $\phi \approx 0.4$ (see Fig. 4.3). Both pulses arrive approximately simultaneously across the different energy bands with small phase shift of the order of $\Delta\phi \lesssim 0.01$ [16]. The flux ratio of the two peaks varies with energy as does the intensity of the bridge component between P1 and P2, which is most prominent in the MeV range and

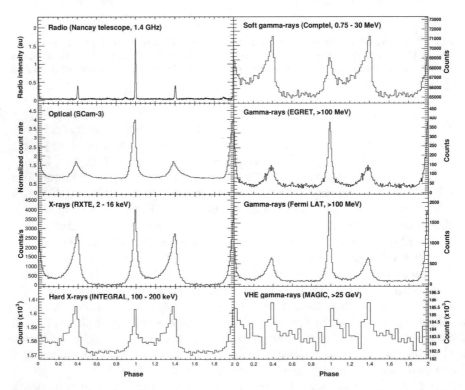

Fig. 4.3 The pulse profile of the Crab pulsar in different energy bands. Two cycles are shown for clarity. The title of each plot gives the energy range and the corresponding instrument. For hard X-rays and soft gamma rays the dominance of the main pulse over the inter-pulse is reversed. Figure taken from Abdo et al. [16]

completely absent in radio at 1.4 GHz. The pulse profile in the radio regime shows a more complex behavior with a precursor to the main pulse at lower frequencies and two high-frequency components trailing the inter-pulse above $\sim$4 GHz [17].

With a rotation period of $P_{Crab} = 33.6$ ms and a slow-down rate of $\dot{P}_{Crab} = 4.2 \times 10^{-13}$ the Crab pulsar is by far the most powerful pulsar in our galaxy [18]. Its spin-down power of $\dot{E} \approx 4.5 \times 10^{38}$ erg s^{-1} is at least an order of magnitude higher than the rest of the pulsars in the Milky Way and is only matched by two pulsars discovered in the Large Magellanic Cloud [19]. Only about 1% of its rotational energy is converted into electromagnetic radiation collimated into beams of light. The spectral energy distribution of this pulsed emission is shown in Fig. 4.2 and is composed of three main components: the first one in the radio band, a second component extending from ultraviolet to soft gamma rays and a third one emerging above energies of approximately 100 MeV [4]. The energetic dominance of the X-ray band with a broad peak at around 10 keV is rather atypical for gamma-ray pulsars. Of all pulsars detected in both X and gamma rays only the Crab and B1509-58 exhibit an equal or greater power of the non-thermal component in the X-ray band as in the hard gamma-ray band [18, 20]. A phase resolved analysis of the spectral index in both bands by Weisskopf et al. [20] using Chandra data and by Abdo et al. [16] using *Fermi*-LAT data revealed a strikingly similar behavior with the smallest index occurring in the middle of the bridge and increasing symmetrically through each peak until reaching maxima in the rising edge of P1 and the falling edge of P2, respectively.

But the most remarkable feature which distinguishes the Crab from all other pulsars lies in its gamma-ray component. As discussed in Sect. 2.4 the vast majority of gamma-ray pulsars seem to exhibit a spectral cut-off at around a few GeV. Therefore it came as a big surprise when MAGIC announced the detection of the Crab pulsar above 25 GeV in 2008 [21], and together with VERITAS even observed pulsed gamma-rays above 50 GeV up to approximately 400 GeV in 2011 and 2012 [22, 23]. So far only two pulsars have been detected at energies above 50 GeV but only the Crab reaches, in a power-law like extension of its spectrum, up to 400 GeV. In the following we will refer to this component as the *very-high-energy tail* of the Crab.

4.2 Observations of the Crab Pulsar at Very High Energies

In 1972 the SAS 2 satellite was the first experiment to record significant pulsed emission from the Crab above $\sim$30 MeV and lead the way in establishing the Crab pulsar as a hard gamma-ray emitter [24]. Due to a failure of the low-voltage power supply it only collected data for less than one year and left the task of getting first detailed information on the high-energy gamma-ray properties of the Crab to the Cosmic Ray Satellite COS-B three years later. With the full COS-B dataset of almost seven years, Clear et al. [25] were able to discover bridge emission and to measure individual spectra for the steady (nebula) and pulsed components (P1, P2 and the bridge) up to $\sim$3 GeV. Following the European COS-B satellite came EGRET on board of NASA'S Compton Gamma Ray Observatory in 1991 that together with COMPTEL provided a complete gamma-ray spectrum of the pulsar from ~1 MeV up to ~10 GeV [26]. The data showed that the high-energy component above ~70 MeV could empirically be described by a power-law with a photon index close to 2 with a hint of a break between 4 and 10 GeV. This was consistent with earlier attempts to detect the pulsar at very-high energies (VHE) from the ground, by for example Lessard et al. [27] with the Whipple telescope or de Naurois et al. [28] with CELESTE, which resulted in stringent upper limits on the flux that required a break or cutoff of the spectrum somewhere above a few GeV.

The very-high-energy tail of the pulsar was detected by two instruments so far: the MAGIC telescopes and VERITAS. Especially the MAGIC collaboration and the Crab pulsar share a long and fruitful history when it comes to publications in peer-reviewed journals, resulting in a total of six articles to date. The first one in early 2008 falls in the mono era of MAGIC and mainly dealt with the continuous VHE gamma-ray emission of the nebula and the performance of the instrument [8]. With a 16 h data-set taken in 2005 they were not able to claim a detection of the pulsar, although they saw a hint of the inter-pulse at a 2.9σ level above their energy threshold of 60 GeV. To verify the hint, a novel trigger system (the so-called *sum trigger*) was developed to lower the threshold down to ~25 GeV [29], and with 22 h of new data taken between October 2007 and February 2008 came the breakthrough. In the *Science Magazine*, Aliu et al. [21] reported the detection of the Crab pulsar for the first time in the VHE regime. The two peaks seen in the pulse profile above 25 GeV are coincident with the ones in the other energy bands, but in contrast to the profile measured by EGRET, where P1 dominates over P2, both peaks have similar amplitudes (see Fig. 4.3). The additional flux measurement at ~25 GeV together with the EGRET spectrum allowed to infer a cut-off at around 20 GeV and to locate the emission region well above the neutron star surface, ruling out the polar-cap (PC) scenario as a valid gamma-ray radiation model for the Crab pulsar.

After the launch of *Fermi* in 2008, the number of known gamma-ray pulsars increased dramatically (see Sect. 2.4). Subsequent population studies by Abdo et al. [18, 30] confirmed the notion that the observed gamma rays are produced rather in the outer part of the pulsar's magnetosphere than close to its surface: (i) most of the light curves show two strong, caustic peaks significantly separated, a pattern typical for outer gap (OG) models; (ii) the substantial fraction of radio faint objects among the gamma-ray pulsars suggests that gamma-ray emission has an appreciably larger extent than the radio beams, such as expected in the outer or slot gap (SG) models; (iii) the PC scenario predicts a super-exponential cut-off in the gamma-ray pulsar spectrum which has not been observed so far. Regarding the Crab pulsar, the *Fermi* LAT could affirm the hint of a break at several GeVs seen by its predecessor EGRET and placed a cut-off energy at around 6 GeV [16].

As a true surprise came the discovery of the very-high-energy tail of the Crab pulsar which was not predicted by any model at that time. In 2011 the VERITAS collaboration announced the detection of pulsed emission above 100 GeV in the *Science Magazine*, followed by more detailed studies from the MAGIC collaboration in the *Astrophysical Journal* in 2011 and in *Astronomy & Astrophysics* in 2012

[22, 23, 32]. The two papers by the MAGIC collaboration concentrated on marginal different but overlapping energy ranges and used different data sets. Aleksić et al. [32] took around 59 h of data with the sum trigger from 2007 until 2009 corresponding still to the mono era of MAGIC and focused on energies below 100 GeV. Aleksić et al. [23] on the other hand analyzed 73 h of data taken after 2009 when the system started operating in stereoscopic mode with a slightly higher energy threshold (see Sect. 3.3).

The pulse profile above 50 GeV again shows two peaks coincident with the ones in the lower energy bands. In contrast to the GeV range, however, the inter-pulse is clearly dominating over the main pulse as can be seen in Fig. 4.4. This fits the trend of an increasing P2 to P1 intensity ratio with energy observed above 100 MeV in the *Fermi*-LAT data. Another difference to the GeV range is the remarkable narrowness of the peaks with full widths at half maximum (FWHM) of approximately 0.027 in phase which is a factor of two to three less than those measured by the *Fermi* LAT at 100 MeV. Transformed into the time domain these phase widths correspond to

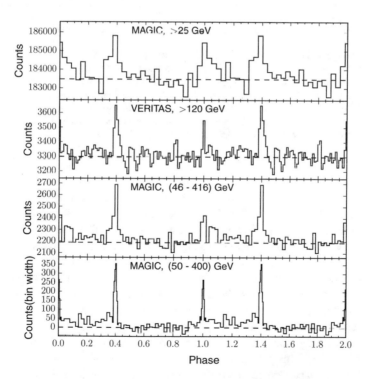

Fig. 4.4 The pulse profile of the Crab pulsar at VHE ordered by its chronological appearance in peer-reviewed journals (see text). The dashed lines denote the respective background levels. At VHE the emission ratio between the main pulse at ∼1.0 and the inter-pulse at phase ∼1.4 is around 0.5. The bridge component is only observed significantly in the bottommost plot where the excess counts are plotted in function of the phase bin width to highlight the sharpness of the peaks. Plots reproduced from (top to bottom): Aliu et al. [21, 22], Aleksić et al. [23, 31]

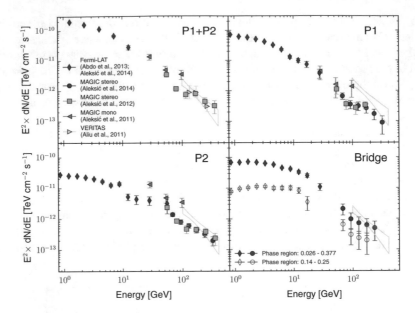

Fig. 4.5 The spectral energy distribution for different components of the Crab pulsar at VHE. The butterfly plot measured by VERITAS for P1 + P2 is plotted for every component as reference. While the *Fermi*-LAT spectral points in the upper left panel are taken from the phase averaged spectrum in Abdo et al. [18], the rest of the panels show the *Fermi*-LAT points obtained by Aleksić et al. [31]. For the bridge component we plot the energy distribution for the two phase regions discussed in the text. MAGIC mono data may have been over-estimated, as discussed in detail in Saito [34]. Plots reproduced from: Abdo et al. [18]; Aleksić et al. [23, 31, 32]; Aliu et al. [22] (Color figure online)

gamma-ray pulses shorter than 1 ms. The energy spectrum reveals a steep power-law like extension after the break at $\sim$10 GeV, the so-called *very-high-energy tail* (see Fig. 4.5). It is well described by a power-law with a spectral index of around 3.5 and extends up to 400 GeV without a hint of a cut-off. Phase-resolved studies by Aleksić et al. [23, 31] showed no significant differences between the spectral indexes of both peaks. The measurement of pulsed emission up to 400 GeV not only constrained the emission site to be at least 10 to 40 stellar radii[1] away from the surface, but also raised doubts about the gamma-ray production mechanism being curvature radiation, as is generally believed (see Sect. 2.4). Only in the extreme case of an accelerating electric field close to the maximum allowed and a curvature radius close to the light-cylinder radius would it be possible to produce gamma rays above 100 GeV [22, 33]. This conclusion motivated several new emission models for the Crab pulsar which advocated inverse Compton scattering as the primary mechanism for very-high-energy gamma rays (see next section).

In 2014 the MAGIC collaboration published another paper on the Crab pulsar reporting the detection of the bridge component above 50 GeV [31]. With 135 h of

---

[1]That is the radius of the neutron star itself, which is around 10 km.

stereo data containing the sample used by Aleksić et al. [23], the study provided the most detailed light curve and phase-resolved spectra of the Crab pulsar at VHE to date by means of deep observations. The increasing ratio of bridge to P1 intensity with energy follows the trend observed above 100 MeV. Originally Fierro et al. [35] defined the bridge region in a relatively narrow phase interval from 0.14 to 0.25 in order to avoid the trailing wing of P1 and the leading wing of P2. The sharp peaks seen in the VHE regime, however, allow to extend the bridge interval considerably to 0.026 and 0.377. Both definitions of the bridge region lead to power-law spectra with indexes of around 3.4, comparable to those of the main and inter-pulse. Aleksić et al. [31] discussed the implications of such bridge emission for current models and concluded that none of them could consistently account for the experimentally observed features in both the energy spectra and light curve of the Crab pulsar.

The latest publication from 2016 was based on the findings of this thesis and will be discussed in detail in the following chapter [36].

4.3   Motivation of the Reanalysis

All previous publications on the Crab pulsar by the MAGIC collaborations had one common theme regarding the analysis: trying to achieve the lowest energy threshold possible. For this reason they only used data taken with small zenith angles where the air mass between the showers and the telescopes is the lowest and therefore the atmospheric transmission for Cherenkov light is the highest (see Chap. 3). The limits were set at $<35°$ in the case of stereo and $<20°$ in the case of mono observations. For the same reason the publications falling into the mono era of MAGIC only considered data taken with the sum trigger[2] which by design had a lower energy threshold than the standard trigger [29]. If, however, we do not require a low energy threshold and shift our interest to the medium and high energy range accessible to MAGIC, a lot more data becomes available which allows us to increase statistics by a factor of more than ~ 2.

We can further boost statistics by simply combining mono and stereo data. One may ask why this was not done in the previous publications. Being mainly interested in the low energy range the reason is twofold: first, the stereoscopic system has a better energy resolution and smaller systematic uncertainties, especially at lower energies (see Fig. 5.2); second, the much higher background rate in the mono data below ~ 300 GeV would significantly worsen the signal-to-noise ratio (see Fig. 4.6, left plot). At lower energies MAGIC stereo performs better than MAGIC mono by a factor of about ten in terms of sensitivity, above a few hundred GeV the improvement is still a factor of three [37]. This is mainly due to the increase in rejection power of hadron-induced air showers by means of stereoscopic image parameters (see Sect. 3.2). In the case of the Crab pulsar above ~ 400 GeV, however, hadronic showers are not the main background component anymore, as is usually the case, but showers

---

[2]Except the very first paper when the sum trigger was still not available [8].

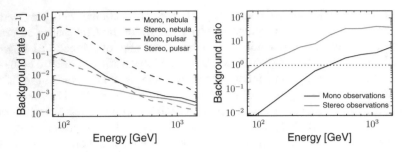

Fig. 4.6 Background rate and ratio versus estimated energy for mono and stereo observations. The mono curves were obtained from the M1 10.6 wobble subsample, the stereo ones from the ST.03.06 < 35° subsample (see Table 5.1). **Left**: In the case of the Crab nebula the difference in the background rate for mono and stereo observations is about a factor 10–50 throughout the energy range. For the pulsar this difference drops to a factor of less than two at energies above 400 GeV. **Right**: We plot the ratio of the two main background components for the Crab pulsar: gamma-ray showers from the nebula and hadronic showers. The gamma-ray showers already dominate the background above ∼100 GeV in the case of stereoscopic observations (Color figure online)

produced by gamma-rays coming from the underlying Crab nebula (see Fig. 4.6, right plot). These showers generate a background that could only be reduced by spatially resolving the pulsar and the nebula, which is still not possible with current generation IACTs. Therefore, the sensitivities of MAGIC mono and MAGIC stereo are comparable when observing the Crab pulsar at energies above ∼400 GeV.

We see that by lifting restrictions on the zenith angle and by combining mono and stereo data we are able to significantly increase our statistics. This allows us to effectively investigate the extension of the very-high-energy tail of the Crab pulsar. What do we expect at energies above 400 GeV? Most of the theoretical models that try to explain the very-high-energy emission from the Crab pulsar are vague regarding the predicted maximum photon energy. Models that solely rely on curvature radiation would have to assume exceptional high accelerating electric fields and curvature radii to produce photons above 400 GeV [33]. Inverse Compton (IC) scattering overcomes these limits but requires some target material. In his modified outer gap model, Hirotani [38] suggests TeV photons through synchrotron self-Compton (SSC) emission from secondary pairs mainly outside the gap (see also [23, 32]). Using a slot gap model Harding and Kalapotharakos [40] also include SSC emission from pair particles in their simulations, but only reach sub-TeV photon energies when assuming an artificial high-energy extension to their cascade pair spectrum. Interestingly they find a very weak SSC component from the primary particles which steeply rises and peaks at a few TeV. This is in concordance with Lyutikov et al. [42], who with a similar but analytical approach, predicts IC emission from the primary beam well into the TeV regime but challenges its detection due to the low flux level. On the other hand Aharonian et al. [39] investigated IC scattering in a cold ultrarelativistic wind at distances between 20 and 70 light cylinder radii away from the pulsar. Their best fit model predicts a clear cut-off at around 400–500 GeV. However, since they lack a physical model explaining the acceleration of the particles in the wind they assume a simple toy model which could be modified to allow for detectable emission above 400 GeV. In somehow more exotic models Chkheidze et al. [43] and Arka and Dubus [44] try to explain the very-high-energy tail by synchrotron emission near the light cylinder during the quasi-linear stage of the cyclotron instability or by a hot plasma in the current sheet of a striped pulsar wind, respectively. In both cases the authors rudimentary outline the possibility of producing photons up to TeV energies. Regarding synchrotron emission from the striped pulsar wind, Mochol and Pétri [41] complemented their wind model to include a synchrotron self-Compton component, which could extend up to tens of TeV in the case of the Crab pulsar. In Fig. 4.7 we compiled some theoretical models where concrete flux predictions were made.

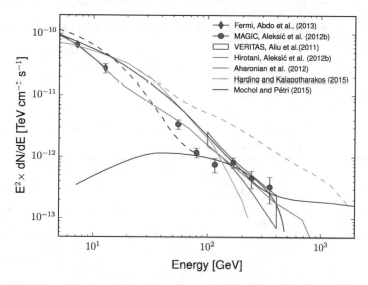

Fig. 4.7 Experimental data (in blue) and theoretical predictions for the very-high-energy tail of the Crab pulsar. While the *Fermi*-LAT spectral points are averaged over the whole phase including the bridge emission, the measurements by Aleksić et al. [23] and Aliu et al. [22] only consider the P1 and P2 components. The model predictions by K. Hirotani (green solid line) are taken from Aleksić et al. [23] (for details on his model the reader is referred to [38]). The red dashed line for Aharonian et al. [39] is the sum of their wind component (solid red line) and an arbitrary model accounting for the emission seen by *Fermi*. Their best fit model assumes that the wind gradually accelerates in a zone 20–50 times the light cylinder away from the pulsar and reaches a maximum Lorentz factor of 10^6. The pair SSC component (solid yellow line) from Harding and Kalapotharakos [40] is plotted for a pair multiplicity (average number of pairs produced by each primary particle) of 3×10^5. It only reaches energies above 1 TeV when an artificial extension to the cascade pair spectrum is assumed (dashed yellow line). The synchrotron and SSC emission from the striped pulsar wind, as advocated by Mochol and Pétri [41], is plotted in violet. We plot their predictions under the assumption of a wind Lorentz factor of 19 and a dissipation distance of 21 light cylinders (for details, see [41]). Plot reproduced from: Abdo et al. [18], Aharonian et al. [39], Aleksić et al. [23], Aliu et al. [22], Harding and Kalapotharakos [40], Mochol and Pétri [41]

In summary, none of the models above identify 400 GeV as a fundamental cut-off in the Crab pulsar spectrum. The maximum photon energy emitted by pulsars is still an open question and at the same time, it is an important ingredient for a better understanding of the emission mechanism at work.

References

1. Green DA, Stephenson FR (2003) The historical supernovae. arXiv, pp 1–12
2. Mayall NU, Oort JH (1942) Further data bearing on the identification of the Crab nebula with the supernova of 1054 A.D. Part II. The astronomical aspects. Publ Astron Soc Pac 54(600):95. https://doi.org/10.1086/125410
3. Trimble V (1973) The distance to the Crab Nebula and NP 0532. Publ Astron Soc Pac 85(October):579. https://doi.org/10.1086/129507
4. Bühler R, Blandford R (2014) The surprising Crab pulsar and its nebula: a review. Rep Progr Phys 77(6):066901. https://doi.org/10.1088/0034-4885/77/6/066901
5. Abdo AA et al (2011) Gamma-ray flares from the Crab Nebula. Science 331(6018):739–742. https://doi.org/10.1126/science.1199705
6. de Jager OC, Harding AK (1992) The expected high-energy to ultra-high-energy gamma-ray spectrum of the Crab Nebula. Astrophys J 396:161. https://doi.org/10.1086/171706
7. Dean AJ et al (2008) Polarized gamma-ray emission from the Crab. Science 321(5893):1183–1185. https://doi.org/10.1126/science.1149056
8. Albert J et al (2008b) VHE γ-ray observation of the Crab Nebula and its pulsar with the MAGIC telescope. Astrophys J 674(2):1037–1055. https://doi.org/10.1086/525270
9. Wilson-Hodge CA et al (2011) When a standard candle flickers. Astrophys J 727(2):L40. https://doi.org/10.1088/2041-8205/727/2/L40
10. Tavani M et al (2009) The AGILE mission. Astron Astrophys 502(3):995–1013. https://doi.org/10.1051/0004-6361/200810527
11. Buehler R et al (2012) Gamma-ray activity in the Crab Nebula: the exceptional flare of 2011 April. Astrophys J 749(1):26. https://doi.org/10.1088/0004-637X/749/1/26
12. Staelin DH, Reifenstein EC (1968) Pulsating radio sources near the Crab Nebula. Science 162(3861):1481–1483. https://doi.org/10.1126/science.162.3861.1481
13. Hewish A et al (1968) Observation of a rapidly pulsating radio source. Nature 217(5130):709–713. https://doi.org/10.1038/217709a0
14. Kuzmin AD (2006) Giant pulses of pulsars radio emission. Chin J Astron Astrophys 6(May):34–40
15. Strader MJ et al (2013) Excess optical enhancement observed with ARCONS for early Crab giant pulses. Astrophys J 779(1):L12. https://doi.org/10.1088/2041-8205/779/1/L12
16. Abdo AA et al (2010a) Fermi large area telescope observations of the Crab Pulsar and Nebula. Astrophys J 708(2):1254–1267. https://doi.org/10.1088/0004-637X/708/2/1254
17. Moffett DA, Hankins TH (1996) Multifrequency radio observations of the Crab pulsar. Astrophys J 468:779. https://doi.org/10.1086/177734
18. Abdo AA et al (2013) The second fermi large area telescope catalog of gamma-ray pulsars. Astrophys J Suppl Ser 208(2):17. https://doi.org/10.1088/0067-0049/208/2/17
19. Manchester RN et al (2005) The Australia telescope national facility pulsar catalogue. Astron J 129(4):1993–2006. https://doi.org/10.1086/428488
20. Weisskopf MC et al (2011) Chandra phase-resolved X-ray spectroscopy of the Crab pulsar. Astron J 743(2):139. https://doi.org/10.1088/0004-637X/743/2/139
21. Aliu E et al (2008a) Observation of pulsed-rays above 25 GeV from the Crab pulsar with MAGIC. Science 322(5905):1221–1224. https://doi.org/10.1126/science.1164718
22. Aliu E et al (2011) Detection of pulsed gamma rays above 100 GeV from the Crab pulsar. Science 334(6052):69–72. https://doi.org/10.1126/science.1208192

23. Aleksić J et al (2012b) Phase-resolved energy spectra of the Crab pulsar in the range of 50–400 GeV measured with the MAGIC telescopes. Astron Astrophys 540:A69. https://doi.org/10.1051/0004-6361/201118166
24. Thompson DJ et al (1977) Final SAS-2 gamma-ray results on sources in the galactic anticenter region. Astrophys J 213:252. https://doi.org/10.1086/155152
25. Clear J et al (1987) A detailed analysis of the high energy gamma-ray emission from the Crab pulsar and nebula. Astron Astrophys 174:85–94
26. Kuiper L et al (2001) The Crab pulsar in the 0.75–30 MeV range as seen by CGRO COMPTEL. Astron Astrophys 378(3):918–935. https://doi.org/10.1051/0004-6361:20011256
27. Lessard RW et al (2000) Search for pulsed TeV gamma-ray emission from the Crab pulsar. Astrophys J 531(2):942–948. https://doi.org/10.1086/308495
28. de Naurois M et al (2002) Measurement of the Crab flux above 60 GeV with the CELESTE cerenkov telescope. Astrophys J 566(1):343–357. https://doi.org/10.1086/337991
29. Aliu E et al (2008b) Observation of pulsed-rays above 25 GeV from the Crab pulsar with MAGIC - SOM. Science 322(5905):1221–1224. https://doi.org/10.1126/science.1164718
30. Abdo AA et al (2010b) The first Fermi large area telescope catalog of gamma-ray pulsars. Astrophys J Suppl Ser 187(2):460–494. https://doi.org/10.1088/0067-0049/187/2/460
31. Aleksić J et al (2014a) Detection of bridge emission above 50 GeV from the Crab pulsar with the MAGIC telescopes. Astron Astrophys 565:L12. https://doi.org/10.1051/0004-6361/201423664
32. Aleksić J et al (2011) Observations of the Crab pulsar between 25 and 100 GeV with the MAGIC I Telescope. Astrophys J 742(1):43. https://doi.org/10.1088/0004-637X/742/1/43
33. Bednarek W (2012) On the origin of sub-TeV gamma-ray pulsed emission from rotating neutron stars. Mon Not R Astron Soc 424(3):2079–2085. https://doi.org/10.1111/j.1365-2966.2012.21354.x
34. Saito T (2010) Study of the high energy gamma-ray emission from the Crab pulsar with the MAGIC telescope and Fermi-LAT. Ph.D. thesis, LMU
35. Fierro JM et al (1998) Phase-resolved studies of the high-energy gamma-ray emission from the Crab, Geminga, and Vela Pulsars. Astrophys J 494(2):734–746. https://doi.org/10.1086/305219
36. Ansoldi S et al (2016) Teraelectronvolt pulsed emission from the Crab pulsar detected by MAGIC. Astron Astrophys 585:A133. https://doi.org/10.1051/0004-6361/201526853
37. Aleksić J et al (2012a) Performance of the MAGIC stereo system obtained with Crab Nebula data. Astropart Phys 35(7):435–448. https://doi.org/10.1016/j.astropartphys.2011.11.007
38. Hirotani K (2015) Three-dimensional non-vacuum pulsar outer-gap model: localized acceleration electric field in the higher altitudes. Astrophys J 798(2):L40. https://doi.org/10.1088/2041-8205/798/2/L40
39. Aharonian FA et al (2012) Abrupt acceleration of a 'cold' ultrarelativistic wind from the Crab pulsar. https://doi.org/10.1038/nature10793
40. Harding AK, Kalapotharakos C (2015) Synchrotron self-compton emission from the Crab and other pulsars. Astrophys J 811(1):63. https://doi.org/10.1088/0004-637X/811/1/63
41. Mochol I, Pétri J (2015) Very high energy emission as a probe of relativistic magnetic reconnection in pulsar winds. Mon Not R Astron Soc Lett 449(1):L51–L55. https://doi.org/10.1093/mnrasl/slv018
42. Lyutikov M et al (2012) The very-high energy emission from pulsars: a case for inverse compton scattering. Astrophys J 754(1):33. https://doi.org/10.1088/0004-637X/754/1/33
43. Chkheidze N et al (2013) On the spectrum of the pulsed gamma-ray emission of the Crab pulsar from 10 MeV to 400 GeV. Astrophys J 773(2):143. https://doi.org/10.1088/0004-637X/773/2/143
44. Arka I, Dubus G (2013) Pulsed high-energy γ-rays from thermal populations in the current sheets of pulsar winds. Astron Astrophys 550:A101. https://doi.org/10.1051/0004-6361/201220110

Chapter 5
Analysis of MAGIC's Data Set of the Crab Pulsar

As the sensitivity of single dish Cherenkov telescopes improves and with the eventual construction of new arrays of ground based Cherenkov telescopes, further observations of pulsar systems should yield definitive data to address the production mechanisms in these systems.

The Whipple Collaboration, 1999

MAGIC observations are programed yearly and grouped into approximately one year long observation cycles, the first one beginning in May 2005. For each cycle members of the MAGIC collaboration submit observation proposals that form the final night-by-night schedule for the telescopes after an evaluation by a committee of experts. The data analyzed in this thesis correspond to Cycles 2–11 and resulted from various observation campaigns with different scientific goals. It is interesting to note that none of the campaigns had the explicit objective of investigating the Crab pulsar spectrum beyond 400 GeV. Particularly in the early cycles and in cycles after major upgrades, a big fraction of the data came actually from technical proposals to monitor and evaluate the performance of the instrument with no immediate scientific purpose.

In the first Cycles 2, 3 and 4, most of the data were taken with the goal of detecting the Crab pulsar at VHE for the first time and to characterize the spectrum in the newly opened energy window. From Cycle 5 on, it was proposed to monitor the Crab nebula to check for variability. Another proposal that significantly contributed to the data set in Cycle 5, 6, 7 and 8 pursued the precise measurement of the Crab nebula spectrum over three decades in energy, in particular the region above 10 TeV [1]. Apart from the monitoring for technical and variability purposes, Cycle 9, 10 and 11 also included special observations of the Crab at very-high zenith angles (above 70°) and observations conducted with new trigger systems. Due to the still unclear performances of these special observations we did not include them in our data set.

© Springer Nature Switzerland AG 2019
D. Carreto Fidalgo, *Revealing the Most Energetic Light from Pulsars and Their Nebulae*, Springer Theses, https://doi.org/10.1007/978-3-030-24194-0_5

The analysis of this unusual big data set was a team effort and done in various steps. In his master thesis[1] the author of this work analyzed the mono subsample and found a first hint of emission above 400 GeV from the Crab pulsar. The hint was double-checked and confirmed by Garrido Terrats [2]. As a continuation of the work started by Gianluca Giavitto,[2] Daniel Galindo reanalyzed all stereo subsamples from 2009 until 2014 and equally found a hint of emission above 400 GeV. The author of this thesis cross-checked and combined the stereo with the mono sample, which ultimately led to a publication of the results in the journal *Astronomy & Astrophysics* [3]. In the following analysis we included some subsamples that were not considered in the publication due to time constraints, namely the M1 9.2, the M1 13.0 and the ST.03.06 subsamples (see Table 5.1).

The first section of this chapter addresses the division of the data set into subsamples while the second section describes the criteria for the quality selection of our data. Section 5.3 explains the data processing and the signal cuts applied. We then show the results of the analysis in form of the pulse profile and the energy spectrum in Sects. 5.4 and 5.5, respectively, before addressing systematic uncertainties and discussing the implication of our findings for pulsar emission models in Sect. 5.6.

5.1 Dividing the Data Set into Subsamples

The data set analyzed for this thesis spans almost 10 years, from February 2007 to March 2016. During this time the MAGIC experiment underwent substantial hardware changes that led to constant improvements in sensitivity as described in Sect. 3.3. In general for each hardware configuration new Monte Carlos (MCs) must be generated that reflect the updated response of the instrument (see Sect. 3.4).

If a change in the instrument response is detected beyond the systematic uncertainty of the experiment, new appropriate MCs are produced and a new *analysis period* is defined. Each MAGIC analysis period has its own set of MCs and receives a tag in the format XX.YY.ZZ:

- *XX* Two letters that indicate if the MCs are valid for monoscopic (M1 or M2) or stereoscopic observations (ST).
- *YY* Two digits that increase for major hardware changes.
- *ZZ* Two digits that increase for minor hardware modifications or weather phenomena that changed the instrument response.

This naming convention became effective when MAGIC started to operate in stereoscopic mode in 2009.

For our mono data taken before 2009, we have three different hardware configurations for the MCs that differ with respect to their optical point spread function (PSF). MAGIC MCs get their optical PSF in two steps. First in the reflector simulation

---

[1]http://www.openthesis.org/document/view/603609_0.pdf, last accessed 15/04/2018.
[2]Gianluca Giavitto was a former Ph.D. student at the Institut de Física d'Altes Energies.

Table 5.1 List of subsamples

Analysis period	Time period	Obs. mode	Zd angle (°)	Obs. time (h)
M1 9.2	Jan 2007–Aug 2009	Wobble	<30	8.4
M1 10.6	Jan 2007–Aug 2009	On/Off	<30	66.1
M1 10.6	Jan 2007–Aug 2009	Wobble	<30	31.2
M1 13.0	Jan 2007–Aug 2009	On/Off	<30	6.5
M1 13.0	Jan 2007–Aug 2009	Wobble	<30	4.8
ST.01.02	Nov 2009–Jun 2011	On/Off	<35	27.5
ST.01.02	Nov 2009–Jun 2011	Wobble	<35	43.1
ST.01.02	Nov 2009–Jun 2011	Wobble	35–50	12.3
ST.02.01	Jan 2012–Feb 2012	Wobble	<35	5.7
ST.02.01	Jan 2012–Feb 2012	Wobble	35–50	1.6
ST.02.03	Mar 2012–May 2012	Wobble	<35	4.9
ST.02.03	Mar 2012–May 2012	Wobble	35–50	5.9
ST.03.01	Sep 2012–Jan 2013	Wobble	<35	27.6
ST.03.01	Sep 2012–Jan 2013	Wobble	35–50	5.4
ST.03.01	Sep 2012–Jan 2013	Wobble	50–62	10.5
ST.03.01	Sep 2012–Jan 2013	Wobble	62–70	5.1
ST.03.02	Jan 2013–Jul 2013	Wobble	<35	21.5
ST.03.02	Jan 2013–Jul 2013	Wobble	35–50	6.5
ST.03.02	Jan 2013– Jul 2013	Wobble	50–62	5.8
ST.03.02	Jan 2013–Jul 2013	Wobble	62–70	3.7
ST.03.03	Jul 2013–Jun 2014	Wobble	<35	27.0
ST.03.03	Jul 2013–Jun 2014	Wobble	35–50	7.8
ST.03.03	Jul 2013–Jun 2014	Wobble	50–62	8.0
ST.03.03	Jul 2013–Jun 2014	Wobble	62–70	5.3
ST.03.06	Nov 2014–Apr 2016	Wobble	<35	31.8
ST.03.06	Nov 2014–Apr 2016	Wobble	35–50	6.7
ST.03.06	Nov 2014–Apr 2016	Wobble	50–62	13.7
ST.03.06	Nov 2014–Apr 2016	Wobble	62–70	9.2

Notes Columns 1 gives the tag of the MAGIC analysis period for each subsample. For the mono data (first 5 rows) we choose the PSF_{MC} as name tag (see text). Column 2 lists the time period corresponding to the analysis period. Columns 3 and 4 list the observation mode and zenith angle distribution of the subsamples. Column 5 gives the effective observation time after quality selection as described in Sect. 5.2. In total we end up with 28 subsamples and a total of 413.6 h of effective observation time

the impact points of the photons in the camera plane are smeared out following a Gaussian distribution with a sigma of 7 mm. Secondly, in order to tune the MCs to match the data, the camera simulation program allows for an additional input PSF_{MC} that gets added in quadrature (for details about the MC production, see Sect. 3.4). Mono MCs were simulated with PSF_{MC} values of 9.2, 10.6 and 13.0 mm. Based on the measurements of the optical PSF of the MAGIC1 telescope obtained from muon

Fig. 5.1 The optical PSF as measured by means of muon runs for the time period when MAGIC was still operating with a single telescope. The red dashed lines define the bins for which the data best matches the corresponding MCs with a PSF$_{MC}$ of 9.2, 10.6 or 13.0 mm. For most of the data we will use MCs with a PSF$_{MC}$ of 10.6 mm

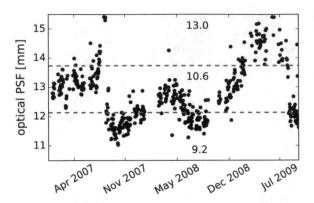

runs since 2007, we divide our mono data into three subsamples that will be analyzed with the best matching set of MCs (see Fig. 5.1).

We not only divide our data set following different hardware configurations but we also have to take into account the observation mode. Normally observations are performed in the *wobble* mode where the source position lies a certain distance away from the center of the camera (for a detailed description of the observation modes, see Sect. 3.4). Hence, the MCs have to be simulated with the same offset. However, especially for our mono data, observations were partially conducted using the *On/Off* mode where the source falls right into the camera center.

We further divide our data with respect to its zenith angle. From the MAGIC site, the Crab is observable from August to April with a zenith angle below 70°. It reaches its highest culmination of around 7° in February. MCs are simulated in fine bins of zenith angle since the instrument response is very sensitive to this parameter. The MARS software automatically takes care of this dependence when calculating, for example, the effective area or the effective observation time (see Sect. 3.4). However, for technical reasons, the signal extraction cuts, namely the *Hadronness* and the *Alpha*/Θ^2 cut, are not determined in function of the zenith angle. Therefore, we divide our data in 4 coarse zenith angle bins: 0°–35°, 35°–50°, 50°–62° and 62°–70°. It should be noted that a common data set in MAGIC consists of far less data and normally spans only one of those zenith bins or a bin with a comparable width.

In total we end up with 28 subsamples as summarized in Table 5.1. For each one of those subsamples we perform an independent analysis and combine the high level products at the end to obtain the final results. Figure 5.2 tries to illustrate the range of instrument responses that our data set covers by plotting the effective area and energy resolution and bias for some representative subsamples.

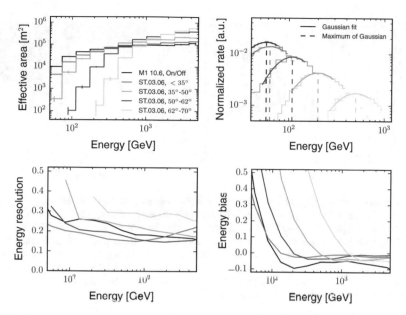

Fig. 5.2 Instrument responses, that is effective area, energy threshold and energy resolution/bias versus true energy, for five representative subsamples. The events were weighted to mimic a power-law like spectrum with an index of −3.42 as measured by Aleksić et al. [4] for the inter-pulse P2 of the Crab pulsar. **Top left**: The effective areas of all subsamples become comparable at around ∼800 GeV. With increasing energy the high zenith angle samples contribute more and more to our total exposure (effective area times observation time). **Top right**: The energy threshold for subsamples with zenith angles above 62° is about ∼500 GeV, but the peak is broad and extends far to lower energies. Therefore one can also obtain scientific results below such defined threshold. **Bottom**: Especially at energies below ∼400 GeV the mono data suffers from a larger energy resolution and bias compared to stereo data with the same zenith angle range. For data taken with higher zenith angles the increasing amount of air mass between the shower and the telescopes leads to a higher uncertainty in the estimation of the energy. Definitions of the energy resolution and bias are given in Sect. 3.4 (Color figure online)

5.2 Quality Selection

MAGIC data can be affected by advert weather conditions or hardware failures. A quality selection aims for removing this data from the analysis to avoid erroneous flux estimations and to improve the sensitivity (see Sect. 3.4).

In a first step we went through all the relevant runbooks written by the shifters during the observation nights and logbooks provided by the MAGIC collaboration to identify data with known hardware failures or very bad weather conditions. In this step we also excluded special observations that were conducted with non-standard wobbble offsets or trigger settings. We then looked at the data itself and at the reports of the subsystems to further discard data of suboptimal or unusual behavior and to discard data taken under moon light. Due to the different hardware configurations

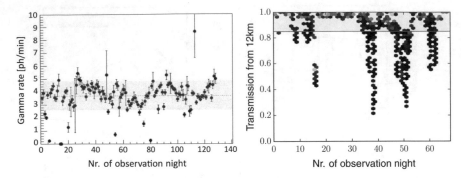

Fig. 5.3 Criteria for the quality selection of the stereo subsamples. **Left**: Nightly gamma rate for data taken below 35° in zenith. The x-axis corresponds to observation nights from the ST.01.02 to ST.03.03 subsamples. Data that is not within errors inside 30% of the mean value (red dashed line and shaded area) is discarded. Plot courtesy of Daniel Galindo. **Right**: Transmission values for the observation nights of the ST.03.06 subsample. The color of the hexagonals denotes the amount of observations time (black less, yellow more). Only data for which the LIDAR showed a transmission of 0.85 or above was selected for the analysis (red shaded area). The transmission was measured from 12 km above the telescopes to the ground (Color figure online)

and improvements of the telescope the criteria for the quality cuts were not the same for all of the subsamples.

For our mono data we concentrated on the event rate and some image parameters following the findings in the master thesis of Ignasi Reichardt Candel.[3] His results point to the root mean square (RMS) of the image parameters *Length* and *Width* as indicators of good quality data, but emphasize that event rate should be the main discriminator. Without any signal extraction cuts this event rate basically corresponds to the rate of hadron-induced air showers. A detailed description of the quality cuts applied to the mono data can be found in the author's master thesis.[4]

For the subsamples from November 2009 until June 2014 the quality selection was performed in a two-step procedure by Daniel Galindo.[5] First, for each analysis period we calculate the average event rate and discard observations that exhibit a rate of more than ±30% of the mean. In a second step we calculate the nightly gamma-rate of the remaining data by extracting the signal from the Crab nebula in three coarse zenith bins. Nights that showed a gamma rate outside 30% of the mean value were discarded as plotted on the left of Fig. 5.3.

For the most recent data corresponding to the analysis period ST.03.06, we based our quality selection on the transmission measurements by the LIDAR (see Sect. 3.3). We discarded all data that showed a transmission of less than 0.85 or for which LIDAR

---

[3]http://citeseerx.ist.psu.edu/viewdoc/download?doi=10.1.1.205.6256&rep=rep1&type=pdf, last accessed 15/04/2018.

[4]http://www.openthesis.org/document/view/603609_0.pdf, last accessed 15/04/2018.

[5]Daniel Galindo is a Ph.D. student at the Universitat de Barcelona at the time of writing.

measurements were not available as depicted on the right in Fig. 5.3. In addition we required the *Cloudiness* to be less than 0.45.

Out of approximately ∼1200 h of Crab data taken since 2007 by the MAGIC telescopes, this selection leaves us with around 414 h of high-quality data. It is worth noting that, to the best knowledge of the author, this is the deepest observation of any object in the VHE regime performed by IACTs to date.

5.3 Data Processing and Signal Cuts

The data was processed using the standard MAGIC analysis package MARS [5]. To benefit from all current features provided by MARS, some of its software had to be modified to cope with our archival mono data. A detailed description of the MAGIC analysis chain can be found in Sect. 3.4.

The signals from the photomultiplier tubes (PMTs) are extracted and calibrated by means of pedestal and calibration runs. We clean the resulting images by using the *sum cleaning* algorithm for most of the stereo data and the *absolute* algorithm for the mono data. In both cases the cleaning parameters are adjusted according to the hardware configuration of the telescopes at the time (see Table D.1 in Appendix D for details). The cleaned shower images are then parametrized by the classical *Hillas* ellipses [6]. Furthermore, in the case of stereo data we can combine the images from both telescopes to obtain powerful stereoscopic parameters [7]. To separate hadronic from gamma-induced air showers we apply the random forest (RF) technique and assign a *Hadronness* parameter to each event. From this point on the analysis proceeds slightly differently depending if we deal with mono or stereo data. For our mono data we use the same RF technique to reconstruct the energy of the event and determine its arrival direction using the *Alpha* parameter (see Sect. 3.4). In contrast, the stereo data relies on simple lookup tables for the energy reconstruction, which are obtained from the MC simulations corresponding to each analysis period. To estimate the arrival direction we use the DISP RF method where the DISP parameters are obtained independently for each telescope by means of a RF and then combined in the stereoscopic image to compute the reconstructed source position. The angular distance between the reconstructed arrival direction and the assumed source position is denoted as Θ.

The last step in our data processing is the assignment of a pulse phase to each event (see Appendix B for details). For this purpose we use the TEMPO2 package initially developed by Hobbs et al. [8] at the Australia National Australia Facility. A plug-in to feed MAGIC data to TEMPO2 was developed by Giavitto [9] and modified by the author of this thesis to function with the cutting edge version of the package. The required ephemerides are obtained from the Jodrell Bank Center of Astrophysics,[6] which has been monitoring the Crab pulsar in radio since January 1987 on a monthly basis [10].

---

[6]http://www.jb.man.ac.uk/~pulsar/crab.html, last accessed 03/03/2018.

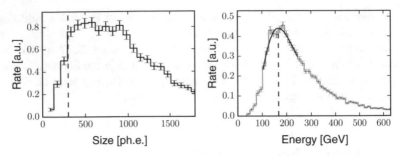

Fig. 5.4 *Size* distribution and energy threshold for a representative mono sample, M1 10.6 On/Off (see Table 5.1). The MC events were reweighed to mimic a power-law like spectrum with an index of −3.42 as measured by Aleksić et al. [4]. **Left**: The *Size* distribution of MC events with a true energy above 400 GeV shows a broad maximum at around ∼600 ph.e. A cut at 300 ph.e. leaves out around ∼6% of the events. **Right**: A Gaussian fit to the true energy distribution in logarithmic space yields a threshold of ∼170 GeV (dashed line) after that size cut

Since we know that our mono samples will spoil the signal quality of the stereo samples at lower energies as shown in Fig. 4.6, we artificially increase their energy thresholds by applying a size cut at 300 ph.e. The value is a compromise between getting rid of the overwhelming background at lower energies and at the same time not loosing any events that could contribute to a signal above 400 GeV. Figure 5.4 illustrates this compromise by plotting the size distribution for MC events with a simulated energy above 400 GeV and the resulting energy threshold after the size cut of 300 ph.e.

All the previous steps leave us with an event list from which we now can extract our signal. Regarding the signal extraction the Crab pulsar is a rather peculiar case for IACTs. For most sources a cut in the *Hadronness* parameter leaves gamma-like showers induced by hadrons as the main background component whose rate decreases rapidly with increasing energy (see the background rate for the Crab nebula in Fig. 4.6). A cut in the *Alpha* or Θ^2 parameter then allows selecting the signal based on the reconstructed arrival direction of the shower. For the Crab pulsar, however, the background at higher energies is dominated by gamma-ray showers from the underlying nebula. To achieve an optimal signal-to-noise ratio in that case, one should aim for loose *Hadronness* cuts that allow a high percentage of MCs to pass. Without any *Hadronness* cut, however, the inclusion of all hadronic showers would drastically increase the background. As a positive side effect, loose *Hadronness* cuts minimize the effect of possible MC-data mismatches. In addition, the periodic signal of the pulsar permits us to extract our signal based on the arrival time of the shower rather then the arrival direction, and hence cuts in *Alpha* and Θ^2 are merely applied to further improve the signal-to-noise ratio (see Sect. 3.4).

Aiming for a detection of the Crab pulsar above 400 GeV we optimize the *Hadronness* and *Alpha/θ^2* cuts following a method discussed in Giavitto [9]. In this method we take advantage of the underlying nebula to estimate the expected signal of the pulsar assuming a given energy spectra. Since the number of excess events is directly proportional to the assumed photon flux of a source we can roughly expect[7] $N_{\text{Pex}} \simeq F_{\text{p}}/F_{\text{n}} \times N_{\text{Nex}}$ excess events from the pulsar, where F_{p} and F_{n} are the assumed integrated photon fluxes for the pulsar and nebula, respectively, and N_{Nex} are the observed excess events from the nebula in our data. For F_{n} and F_{p} we take the results from Aleksić et al. [1, 4],[8] respectively, and assume a continuation of the pulsar power-law spectrum beyond 400 GeV. The expected significance of our pulsed signal can then be written as:

$$S_{\text{p}} = S_{Li\&Ma}\left(N_{\text{on}} \times \Gamma_{\text{TP}} + N_{\text{Pex}}, N_{\text{on}} \times \Gamma_{\text{OP}}, \frac{\Gamma_{\text{TP}}}{\Gamma_{\text{OP}}}\right) \tag{5.1}$$

$S_{Li\&Ma}$ (ON, OFF, α)	The significance of a signal obtained by a counting experiment following Eq. 17 in Li and Ma [11]. ON is the number of events in a region where we suspect a signal. OFF is the number of events in a background region without any signal. α is a normalization factor between ON and OFF.
N_{on}	The total number of events coming from the direction of the Crab nebula after all signal extraction cuts (including the remaining hadronic background).
N_{Pex}	Expected number of excess events from the pulsar after all signal extraction cuts (see text above).
Γ_{TP}	The total phase width of the On phase region of the Crab pulsar. Here we the sum up the two phase widths defined by Aleksić et al. [4] for P1 and P2, that is $0.043 + 0.045$.
Γ_{OP}	The width of the Off phase region of the Crab pulsar, that is 0.35 (adopted from Aleksić et al. [4]).

With this definition of S_{p} we are now able to estimate our expected significance in function of the applied *Hadronness* and *Alpha/Θ^2* cuts. To obtain the optimal cut values we perform a scan in the *Hadronness-Alpha/Θ^2* plane and look for the maximum of S_{p} (see Fig. 5.5). This scan is performed for each subsample individually to take into account the different analysis periods and zenith angles of the observations. For the scans the data is binned in a *low energy* bin, from 100 to 400 GeV, and a *high energy* bin with events above 400 GeV. The idea behind the low energy bin is to cross-check our signal with previous publications by MAGIC. Table 5.2 summarizes the resulting cut values and their respective cut efficiencies, that is the percentage of MC events that survive the corresponding cut assuming a power law spectrum with index -3.57 [4].

These optimized cuts are used to extract the events for the pulsar light curves presented in the next section. For the spectra discussed in Sect. 5.5 we follow a different approach according to the standard analysis procedure of MAGIC (see Sect. 3.4). Since

---

[7]The cut efficiencies of the signal extraction cuts and the instrument response can be slightly different for the Crab nebula and pulsar.

[8]For the case where the on phase regions are defined as 0.983–0.026 and 0.377–0.422 for P1 and P2, respectively (see [4], for details).

(100 - 400) GeV

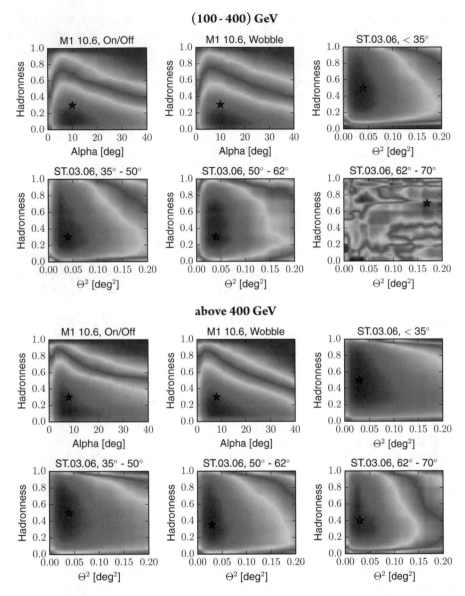

Fig. 5.5 Cut optimization scans for six representative subsamples indicated in the title of each plot. The top two rows correspond to the low energy bin from 100 to 400 GeV, and the bottom two to the energy bin above 400 GeV. The color scale depicts the simulated significance S_p (see text) and corresponds to different absolute values of S_p in each plot (values increase from dark blue to dark red). In addition to looking for the maximum of S_p we impose the following restrictions to the final cut values shown as blue stars to prevent too low cut efficiencies: $Alpha > 8°$, $\Theta^2 > 0.01$ deg^2 and $Hadronness > 0.3$. Because the energy threshold for the 62°–70° subsamples is around ~500 GeV (see Fig. 5.2) the cut optimization does not work properly in the low energy bin for these subsamples (Color figure online)

Table 5.2 Signal extraction cuts for the pulsar light curve

Subsample	(100–400) GeV				Above 400 GeV			
	α/Θ^2	H	eff_α	eff_H	α/Θ^2	H	eff_α	eff_H
M1 9.2, Wobble, <30	10	0.30	81	83	8	0.30	87	89
M1 10.6, Wobble, <30	10	0.30	80	84	8	0.30	88	93
M1 10.6, On/Off, <30	10	0.30	82	85	8	0.30	89	91
M1 13.0, On/Off, <30	8	0.40	76	87	8	0.35	89	93
M1 13.0, Wobble, <30	8	0.30	73	82	8	0.30	87	90
ST.01.02, On/Off, 5–35	0.03	0.60	80	97	0.03	0.45	95	95
ST.01.02, Wobble, 5–35	0.04	0.60	85	97	0.03	0.60	93	98
ST.01.02, Wobble, 35–50	0.04	0.40	78	92	0.03	0.45	90	97
ST.02.01, Wobble, 5–35	0.03	0.75	79	99	0.03	0.60	92	98
ST.02.01, Wobble, 35–50	0.05	0.60	83	97	0.03	0.80	89	100
ST.02.03, Wobble, 5–35	0.04	0.55	86	96	0.03	0.65	93	98
ST.02.03, Wobble, 35–50	0.05	0.40	83	92	0.03	0.65	89	99
ST.03.01, Wobble, 5–35	0.03	0.60	80	97	0.03	0.60	92	98
ST.03.01, Wobble, 35–50	0.04	0.40	80	93	0.03	0.50	89	97
ST.03.01, Wobble, 50–62	0.04	0.40	75	93	0.03	0.55	83	98
ST.03.01, Wobble, 62–70	0.08	0.80	79	95	0.03	0.35	74	95
ST.03.02, Wobble, 5–35	0.03	0.55	81	96	0.03	0.55	91	97
ST.03.02, Wobble, 35–50	0.04	0.30	79	88	0.03	0.55	89	98
ST.03.02, Wobble, 50–62	0.04	0.30	75	90	0.03	0.70	84	99
ST.03.02, Wobble, 62–70	0.01	0.50	22	88	0.04	0.60	81	99
ST.03.03, Wobble, 5–35	0.03	0.50	81	96	0.03	0.60	91	98
ST.03.03, Wobble, 35–50	0.04	0.40	80	93	0.03	0.60	88	98
ST.03.03, Wobble, 50–62	0.04	0.35	75	92	0.03	0.55	84	99
ST.03.03, Wobble, 62–70	0.06	0.95	75	99	0.03	0.45	75	98
ST.03.06, Wobble, 5–35	0.04	0.50	85	95	0.03	0.50	92	97
ST.03.06, Wobble, 35–50	0.04	0.30	78	88	0.04	0.50	90	97
ST.03.06, Wobble, 50–62	0.04	0.30	73	88	0.03	0.35	82	95
ST.03.06, Wobble, 62–70	0.17	0.70	88	95	0.03	0.40	72	94

Notes The subsample names in the first column are composed of the analysis period, the observation mode and the zenith angle of the observation (see Table 5.1). Columns 2 to 5 correspond to the low energy bin from 100 to 400 GeV, Columns 6 to 9 to the energy bin above 400 GeV. Columns 2 and 6 give the cut values for *Alpha* (α) and Θ^2 in case of monoscopic and stereoscopic data, respectively. Columns 3 and 7 give the cut values for the *Hadronness* cut. The cut efficiencies for the α/Θ^2 and *Hadronness* cuts in Columns 4, 5, 8 and 9 are given in percentage

the performance of the *Hadronness* parameter and the angular resolution depend on the energy of the events we apply different cuts in each energy bin of our spectrum. The cut values are determined by fixing the efficiency of the cuts to 75% for the $Alpha/\Theta^2$ and 95% for the *Hadronness* parameter. As discussed before, in the case of a gamma-ray dominated background one should opt for a high cut efficiency to achieve an optimal signal-to-noise ratio. A high cut efficiency also minimizes the effect of potential mismatches between MCs and data, which becomes crucial when calculating the effective area for the flux estimation. These considerations are reflected in the choice of 95% for the *Hadronness* cut efficiency, slightly higher compared to the conventional value of 90% in the standard analysis chain of MAGIC. In principle the same considerations also apply for the $Alpha/\Theta^2$ cut. However, events with a high $Alpha/\Theta^2$ value tend to be badly reconstructed and can spoil the energy resolution. For this reason we stick to the general value of 75% of the standard analysis chain. We note that instead of fixing the cut efficiencies, we alternatively could determine the cut values by maximizing the expected significance in each of the energy bins of the spectrum, as discussed in the previous paragraph, and impose the restrictions to the cut efficiencies in the maximization. In this case one should apply some kind of smoothing to the cuts in the end to ensure a reasonable cut evolution between neighboring energy bins and to avoid selection biases when statistics are low, as done in Aleksić et al. [4]. Both approaches should give compatible results.

5.4   Pulsar Light Curve Below and Above 400 GeV

Figure 5.6 shows the pulsar light curves in the energy bin from 100 GeV to 400 GeV and above 400 GeV after applying the signal extraction cuts. To evaluate the significance of the signal we use Eq. 17 in Li and Ma [11] and choose a priori the signal and background phase regions following [4]. From a Gaussian fit to both peaks in the energy range from 46 to 416 GeV, they defined the signal phase regions as the peak positions $\pm 2\sigma$ resulting in the phase intervals [0.983, 1.026] for P1 and [0.377, 0.422] for P2 (see shaded regions in Fig. 5.6). The background level was estimated in the interval [0.52, 0.87] based on the publication by Fierro et al. [13].

In the energy bin from 100 to 400 GeV we clearly see both peaks emerging from the strong background. The numbers of excesses events found in both phase regions are 645 ± 135 for P1 and 1163 ± 140 for P2, corresponding to a significance of 4.8σ and 8.5σ, respectively, and an overall significance of 9.0σ considering both peaks. The peak emission ratio is $P1/P2 = 0.55 \pm 0.13$. Compared to previous MAGIC publications by Aleksić et al. [4, 14] this signal strength seems rather weak considering the much longer observation time of our data set by a factor of about 4. This is expected, however, since the background of the mono and high zenith samples dilutes the signal of the low zenith stereo samples at lower energies (see Sect. 4.3). If we only consider stereo data with zenith angles below 35°, which identifies the subsamples with the lowest energy threshold and the best gamma-hadron separation

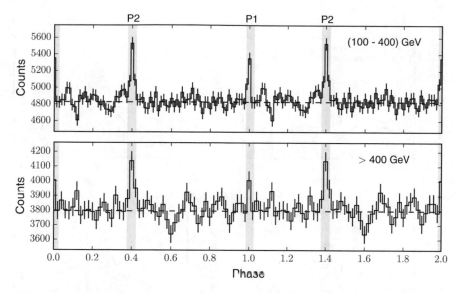

Fig. 5.6 Folded light curves of the Crab pulsar in the energy bins from 100 to 400 GeV and above 400 GeV obtained with the MAGIC telescopes after ~414 h of effective observation time. Two cycles are shown for clarity. The shaded regions denote the signal phase regions from where we take our On counts for P1 and P2. The background level is estimated in the phase interval [0.52, 0.87] and is drawn as a horizontal dashed line. The excess found in the P2 phase region for energies above 400 GeV yields a significance of 5.2σ

power, we end up with 189 h of observation time and and an overall significance for P1 + P2 of 9.8σ. This sensitivity is comparable to the one of Aleksić et al. [4].

To cross-check the shape of the peaks with previous findings, we fit our pulse profile with a double Gaussian plus a constant maximizing an unbinned likelihood function as shown in Fig. 5.7 (see Appendix D for details). The background level found from the fit agrees well within statistical uncertainties with the average number of events found in the phase interval [0.52, 0.87]. Including a constant term to account for a possible contribution of the bridge emission between 0.026 and 0.377 in phase does not significantly improve our likelihood. We also look for possible asymmetries in the peak shapes by means of asymmetric Gaussians, and find that P1 exhibits a sharper fall than rise at a 2.2σ level (see Table 5.3). The same behavior is seen for P2 but only with a significance of 1.5σ. As a last check we fit the two peaks with Lorentz functions that allow for potential wider tails, but obtain lower likelihood values as compared to the Gaussian case. However, since Lorentz and Gauss functions are not nested models the difference in their likelihoods cannot be easily quantified statistically. When fitting a very fine-binned pulse profile maximizing a Poissonian likelihood we cannot exclude either shape by its χ^2 value.

The light curve for energies above 400 GeV reveals pulsed emission from the inter-pulse at a 5.2σ level. The main pulse is barley visible at these energies and yields a significance of only 2.0σ. With 188 ± 97 excess counts in the P1 phase region and

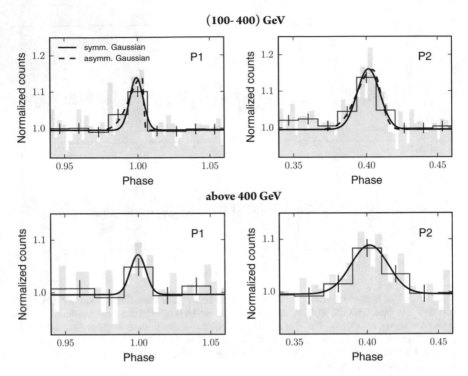

Fig. 5.7 Unbinned likelihood fits to the pulse profile of the Crab in the energy bin from 100 GeV to 400 GeV and above 400 GeV. The step histograms exhibit the same binning as in Fig. 5.6, the shaded histograms have a binning 5 times finer. Below 400 GeV the signal strength allows to test for possible asymmetries in the peak shapes (see Table 5.3). For P1 a sharper fall than rise is preferred at a 2.2σ level. At energies above 400 GeV a simple Gaussian should be sufficient to describe our peaks

Table 5.3 Fit results of the pulse profiles below and above 400 GeV

Energy range (GeV)	Peak	μ_s ($\times 10^{-3}$)	FWHM ($\times 10^{-3}$)	μ_a ($\times 10^{-3}$)	HW$_r$ ($\times 10^{-3}$)	HW$_f$ ($\times 10^{-3}$)
100–400	P1	$999.2^{+1.2}_{-1.4}$	$10.7^{+3.5}_{-2.4}$	$3.4^{+1.7}_{-2.2}$	$10.5^{+4.7}_{-3.6}$	$0.8^{+2.7}_{-0.8}$
	P2	$401.4^{+1.4}_{-1.5}$	$18.3^{+3.7}_{-3.1}$	$404.3^{+1.6}_{-1.9}$	$12.7^{+5.2}_{-3.6}$	$6.5^{+2.9}_{-2.7}$
>400	P1	$999.7^{+3.8}_{-3.4}$	$12.5^{+9.7}_{-6.7}$	–	–	–
	P2	$401.8^{+4.6}_{-4.3}$	$30.3^{+12}_{-8.8}$	–	–	–

Notes Column 3 gives the peak position when fitting the pulse profile with a symmetric Gaussian, while Column 4 states the corresponding *full width at half maximum* (FWHM). Column 5 gives the peak position as obtained from an asymmetric Gaussian fit, Columns 6 and 7 contain the corresponding *half width at half maximum* (HW) of the *rising* and *falling* edges. The given errors are purely statistical and were taken from the MINOS processor of Minuit that we use to maximize our likelihood (see Appendix D). Systematic errors can arise due to the limited accuracy of the ephemerides used to determine the rotation phases (see text). For the position of the peaks we estimate this error to be ±0.0039. For the FWHMs it is −0.0052(−0.0025) for P1(P2) at energies below 400 GeV, and −0.0040(−0.0014) for energies above 400 GeV. Fit results in the energy range from 100 to 400 GeV are in good agreement with [4, 12], except for the width of P1 (see text)

519 ± 101 in P2, the emission ratio between the two peaks is $P1/P2 = 0.36 \pm 0.20$.
We also apply the *H-test* by de Jager et al. [15], which does not make any assumption
on the peak positions, and obtain a significance of 3.4σ that pulsed emission is
present. Due to the finite energy resolution of the instrument we inevitably will get
some spillover of events with a true energy below 400 GeV that cannot be corrected
for on an event-by-event basis in our pulsar light curves. We estimate this effect by
comparing the true and estimated energies in our MCs of each subsample and find that
around $\sim16\%$ of our gamma-ray events coming from P2 have a true energy lower
than 400 GeV. For energies above 500 GeV the excess in the P2 phase interval still
yields a significance of 4.4σ and the spillover of events lower than 400 GeV drops
to $\sim7\%$ (see Appendix D for details on the estimation).

When fitting both pulses with two symmetric Gaussians the resulting peak positions coincide with the ones in the energy bin from 100 to 400 GeV. On the other
hand, the width of P2, measured as the *full width at half maximum* (FWHM) of its
Gaussian, seems to become broader. We find this effect to be at a 2σ level by fixing
the sigma of P2's Gaussian to the value obtained in the (100–400) GeV bin and
comparing the resulting likelihood values. This is rather unexpected since it breaks
the trend of the narrowing of the pulse with energy as seen in the MeV to ~100 GeV
range (see Fig. 3 of Aleksić et al. [4]). We also try to employ asymmetric Gaussians
and Lorentz functions in the fit but get very similar likelihood values. Hence, the
conservative approach of using a Gaussian parametrization should be sufficient to
describe our peaks at energies above 400 GeV. We note that in both energy ranges
we measure a P1 peak width substantially narrower than the one reported by Aleksić
et al. [4] from where we adopted our signal phase intervals a priori.[9] If we choose
the intervals following [12], whose fit results are in very good agreement to ours,
the significance of P1 is 6.6σ and 3.0σ below and above 400 GeV, respectively.
However, to avoid any potential selection bias we will stick to our a priori defined
phase intervals with the disadvantage of a possible higher noise contribution to the
signal.

Uncertainties in the ephemerides can lead to systematic errors in the peak positions and widths that are not taken into account during our fits of the pulse profiles.
We estimate these errors from the accuracy provided by the Jodrell Bank Center of
Astrophysics with which their monthly ephemeris is believed to describe the observed arrival times (TOAs) during the whole calender month. Since we used many
ephemerides for our 9 year-long data set we average over the quoted accuracies after
transforming them from time to phase units and obtain a value of 0.0039 which corresponds to ~132 μs. This value directly reflects the precision with which we can
determine the peak position. Uncertainties in the TOAs can only translate to actual
narrower peaks than the ones observed in our light curves. To estimate this error we
look for the width of a Gaussian that folded with a Gaussian of $\sigma = 0.0039$ gives us
the observed width of the peak in the light curve. This results in a systematic error of
$-0.0053(-0.0025)$ for the FWHM of P1(P2) below 400 GeV, and $-0.0040(-0.0014)$
for energies above 400 GeV.

---

[9]As discussed in Aleksić et al. [4] their fit to P1 could be affected by the bridge emission.

Table 5.4 Parametrization of the Crab pulsar spectra

Energy range (GeV)	Peak	E_0 (GeV)	f_{E_0} (cm^{-2}s^{-1}TeV^{-1}]	α	χ^2/d.f.
>85	P1	120	$(2.14 \pm 0.38) \times 10^{-11}$	3.28 ± 0.46	1.2/4
	P2	120	$(4.59 \pm 0.40) \times 10^{-11}$	3.13 ± 0.18	10.1/8
>10	P1	50	$(4.74 \pm 0.50) \times 10^{-10}$	3.54 ± 0.09	2.7/6
	P2	50	$(6.19 \pm 0.49) \times 10^{-10}$	3.01 ± 0.06	12.4/9

Notes Fit results to our MAGIC data above 85 GeV (Fig. 5.8) and *Fermi*-LAT plus MAGIC data above 10 GeV (Fig. 5.9). While the results above 85 GeV are obtained by a forward unfolding method, above 10 GeV we perform a correlated χ^2-fit to *Fermi*-LAT spectral points and the unfolded MAGIC points (see text). In both cases we assume a power-law function (PWL) for the flux $dN/dE = f_{E_0}(E/E_0)^{-\alpha}$. For the forward unfolding the number of degrees of freedom is taken from the distribution of excess events in estimated energies, which may deviate from the number of unfolded spectral points (see Sect. 3.4)

5.5 Energy Spectrum up to 1 TeV

We calculate the spectral energy distribution (SED) for P1 and P2 separately as shown in Fig. 5.8. The spectral points are unfolded by means of the Bertero method, and thus the statistical errors are correlated[10] reflecting our energy resolution and bias of ~20 to 40% depending on the energy. Instead of performing a correlated fit to the spectral points we determine the spectral shape and parameters by applying a forward folding, which is a robuster method to parametrize the data (see Sect. 3.4 and [16], for details). Both spectra are well described by power-law functions (PWL) $dN/dE = f_{E_0}(E/E_0)^{-\alpha}$ with photon indexes α of around 3.2 resulting in reduced χ^2 values of 1.2/4 and 10.1/8 for P1 and P2, respectively (see Table 5.4). The results of the forward folding calculated with a normalization energy of 120 GeV, which is close the the respective decorrelation energy of 116 GeV for P1 and 134 GeV for P2, are in agreement with previous results obtained by Aleksić et al. [4, 14]. We also check the relative errors of the spectral parameters at the exact decorrelation energy for each peak but find only negligible differences of less than 1%. The ratio of the normalization constants of P1 and P2 at 120 GeV is 0.47 ± 0.09, which is consistent with the values directly derived from the light curves below and above 400 GeV. In the case of the inter-pulse the power-law spectrum extends up to approximately ~1.2 TeV showing for the first time evidence of pulsed emission above 1 TeV. The main pulse cannot be measured beyond ~500 GeV where its flux drops to almost 0.1% of the flux from the nebula. For energies where the significance of the excess before the unfolding falls below 1.5σ we compute upper limits (ULs) to the differential flux at the 95% confidence level following [17] and under the assumption of the power-law spectrum found in this work. A 20% change in the assumed photon index yields a variation of less than 15% in the ULs. Our ULs do not constrain any possible cut-off in the spectra given the current sensitivity of the instrument.

---

[10]We provide the full covariance matrix in Appendix D.

To quantify the exclusion of a cut-off below 1 TeV in our inter-pulse spectrum we repeat the forward folding under the assumption of a power-law with an exponential cut-off (EPWL) following the expression $dN/dE = f_{E_0}(E/E_0)^{-\alpha}\exp(-E/E_c)$. Fixing the cut-off energy E_c and comparing the resulting χ^2 values in both cases, we can exclude a EPWL in favor of a simple PWL at 90 and 95% confidence levels for cut-offs at 1.13 and 0.77 TeV, respectively (see Fig. 5.8).

The extrapolation of our spectra to lower energies agrees within statistical errors with the spectral points measured by *Fermi*-LAT above 10 GeV, which already show a deviation from the expected exponential cut-off usually observed in gamma-ray pulsars (see dashed lines in Fig. 5.9). Therefore, we perform a joint χ^2-fit to our MAGIC and *Fermi*-LAT spectral points above 10 GeV in which we incorporate the correlations of the MAGIC errors due to the unfolding procedure and assume no correlations between the *Fermi*-LAT points. The spectral component of the pulsar above 10 GeV can be described satisfactorily well by simple power-law functions that yield reduced χ^2 values of 2.7/6 and 12.4/9 for P1 and P2, respectively (see Table 5.4). The normalization energy for the fits was fixed to 50 GeV, close to the respective decorrelation energy of 33 GeV for P1 and 90 GeV for P2. The significant difference in the photon indexes of P1, 3.54 ± 0.09, and P2, 3.01 ± 0.06, indicates that the intensity of P1 drops more rapidly with energy than for P2, which corroborates the trend of a decreasing intensity ratio $P1/P2$ seen by *Fermi*-LAT above 1 GeV.

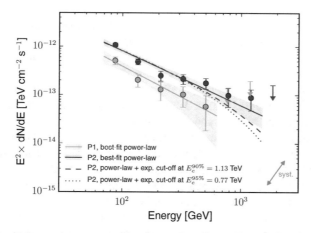

Fig. 5.8 Spectral measurements of the main (blue) and inter-pulse (red) after 414 h of observation time by the MAGIC telescopes. The spectral points are obtained by unfolding the data using the Bertero method. Our measurements are parametrized by means of a forward unfolding method assuming a power-law as indicated by the straight lines. The corresponding butterflies consider statistical errors only. The results of the forward unfolding are given in Table 5.4. The most conservative systematical errors on the energy scale and flux normalization are depicted by the grey arrow in the bottom right corner (see text). The dashed and dotted lines show the best fit results when assuming a power-law with an exponential cut-off at $E_c = 1.13$ TeV and $E_c = 0.77$ TeV, respectively. The numerical values for the spectral points are given in Table D.2 of Appendix D (Color figure online)

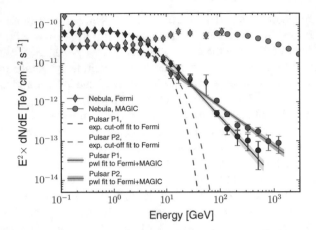

Fig. 5.9 Spectral measurements of the Crab pulsar (blue and red) and nebula (yellow) in the energy range above 100 MeV by *Fermi*-LAT (diamonds [1]) and the MAGIC telescopes (filled circles, this work). Both pulsar analyses, *Fermi* and MAGIC, were conducted in the same phase intervals that are discussed in the beginning of Sect. 5.4. The dashed lines parametrize only the *Fermi*-LAT data with a power-law plus exponential cut-off as stated in Aleksić et al. [1]. The joint *Fermi* and MAGIC fits include spectral measurements above 10 GeV and are shown as straight lines. The corresponding butterflies consider statistical errors only. The results of the joint fits are given in Table 5.4 (Color figure online)

We note that we do not consider systematical discrepancies in the flux estimations between the two instruments in the fit, however, any differences should affect both peak spectra in the same way. From the fits we estimate the turnover, where P2 becomes intenser than P1, to lie between 15 and 44 GeV.

To ensure a good understanding of all possible systematic errors, we furthermore compute the Crab nebula spectrum from the same data set, in the same energy range with the same energy binning and using the same signal extraction cuts as for the pulsar analysis, but excluding the subsamples that were taken in On/Off mode, which constitute approximately 24% of our whole set. We exclude these subsamples to avoid systematic effects caused by an inadequate choice of off data that would only affect the nebula analysis but not the pulsar one. The nebula spectrum is consistent with the most recent published Crab nebula analysis by MAGIC in Aleksić et al. [1] and seems to connect smoothly to the *Fermi*-LAT data from Aleksić et al. [14], which confirms a good performance of our spectral analysis in the energy range above ∼85 GeV (see Fig. 5.9). The systematic errors corresponding to our data set were evaluated in three publications by the MAGIC collaboration, each one dealing with a different major analysis period (Mono: [18]; ST.01: [4]; ST.03: [19]).[11] The most conservative estimates from all three publications are a systematic uncertainty of 17% on the energy scale and 19% on the flux normalization at energies below ∼100 GeV, as indicated by the grey arrow in Fig. 5.8. The arrow also takes into account the

---

[11] In all three papers the systematic errors were only studied for data taken below 45° in zenith. A dedicated study for observations at higher zenith angles is still ongoing.

effect of the uncertainty in the energy scale on the flux assuming a spectral index of 3.28. At medium energies, above $\sim$300 GeV, the energy scale is determined with a precision of about 16% in the case of mono data and around 15% for data taken in stereoscopic mode. Due to the low signal to background ratio of our data and the resulting uncertainties in the unfolding procedure we estimate our systematic error on the photon index to be 0.3, which is higher than the 0.2 and 0.15 for mono and stereo data, respectively, stated in the MAGIC performance papers. To study the effect of the uncertainty on the energy scale for the exclusion of a cut-off below 1 TeV, we repeat the last step of the analysis changing the estimated energy by $\pm16\%$ and $\pm15\%$ for our mono and stereo data, respectively,[12] and again perform the forward unfolding under the assumption of a EPWL. Note that this does not produce a simple shift of the spectrum along the energy axis, but also changes its hardness. This variation in the whole data set is a rather conservative approach since it assumes the unlikely case that throughout our 414 h of observation the *average* Cherenkov light yield was overestimated or underestimated by 16 and 15%, respectively, relative to the MCs. The resulting errors for the lower limits at the 90 and 95% confidence levels are $E_c^{90\%} = 1.13 \, ^{+0.20}_{-0.45}$ TeV and $E_c^{95\%} = 0.77 \, ^{+0.14}_{-0.30}$ TeV.

5.6 Discussion and Conclusions

With 414 h of archival and recent data taken by the MAGIC telescopes, the analysis in this work constitutes the deepest observation of the Crab pulsar in the very-high-energy (VHE) range to date. We found pulsed VHE emission from the inter-pulse above 400 GeV with a significance of 5.2σ that allows us to compute a spectrum in an unprecedentedly broad energy range from $\sim$85 to $\sim$1.2 TeV and provides, for the first time, evidence of pulsed emission above 1 TeV. Our spectrum can be parametrized well by a simple power-law function with spectral index $\alpha = 3.13 \pm 0.18_{\text{stat}} \pm 0.3_{\text{syst}}$. The power-law is preferred over an exponential cut-off at 90% and 95% confidence levels for cut-off energies at $E_c^{90\%} = 1.13 \, ^{+0.20}_{-0.45}$ TeV and $E_c^{95\%} = 0.77 \, ^{+0.14}_{-0.30}$ TeV, respectively. The spectrum of the main pulse can also be described by a power-law with $\alpha = 3.28 \pm 0.46_{\text{stat}} \pm 0.3_{\text{syst}}$ but is not detected beyond $\sim$500 GeV where its flux drops to almost 0.1% of the flux from the Crab nebula.

The pulse shape of P1 in the energy range from 100 to 400 GeV exhibits a slight asymmetry, that is a sharper fall than rise, at a 2.2σ level, which is also seen in the infrared to ultraviolet band and by *Fermi*-LAT at energies below 1 GeV [20, 21]. At energies above 400 GeV the signal of the pulses is too weak to discriminate between more complicated pulse shapes and conservative Gaussian parametrizations are sufficient to describe both of the peaks. The inter-pulse P2 clearly dominates the

---

[12]This method provides an intermediate solution to estimate the effect, since a proper study should incorporate the uncertainty on the light yield directly in the MC simulation. For 28 different sets of MCs, however, such an exercise would imply an unrealistic amount of work and is out of scope for this thesis.

pulsar light curve above 400 GeV with an intensity ratio of $P1/P2 = 0.36 \pm 0.20$. This result seems to follow the trend of a decreasing $P1/P2$ ratio with energy seen by *Fermi*-LAT above 1 GeV and is corroborated by the joint fits of *Fermi*-LAT and MAGIC measurements (see Fig. 5.9), which yield significantly different photon indexes of 3.54 ± 0.09 and 3.01 ± 0.06 for P1 and P2, respectively.

Pulsed gamma-rays with an energy of ~ 1 TeV lead to implications for the site of the emission region and for the emission mechanism in the Crab pulsar.

Location of the emission region

Photons with an energy above ε_{max}, which are emitted near the neutron star at a distance r, cannot escape due to magnetic pair creation and photon splitting caused by the strong magnetic field. The cut-off energy ε_{max} in this scenario can be estimated by Baring [22]

$$\varepsilon_{max} \simeq 0.4\sqrt{P} \left(\frac{r}{R_0}\right)^{1/2} \max\left\{1, \frac{0.1 B_{cr}}{B_0}^3\right\} \text{GeV}, \qquad (5.2)$$

where P, R_0 and B_0 are the period, radius and magnetic field strength at the surface of the pulsar, respectively, and $B_{cr} = 4.4 \times 10^{13}$ G is the critical magnetic field at which the cyclotron energy equals the electron rest mass [23]. Hence, TeV photons from the Crab pulsar, for which $P = 0.0336$ s, $R_0 = 10$ km and $B_0 = 3.8 \times 10^{12}$ G, must be produced at least 15 stellar radii away from the neutron star's surface. Lee et al. [24] argue that [22] actually underestimates the characteristic absorption length and by means of a numerical approach predict an escape radius of $r \gtrsim 60$ stellar radii for TeV photons. These lower bounds challenge some of the *slot gap* and *annular gap* models that place the gamma-ray emission site at a few and several tens of stellar radii, respectively, above the pulsar's surface [25, 26].

Emission mechanism

For the gamma-ray emission mechanism, on the other hand, we can provide constraints to the standard scenario, in which gamma-rays are produced via synchro-curvature radiation. Following [27] we consider monoenergetic electrons with elementary charge e and a Lorentz factor γ that move along a magnetic field line with curvature radius R_c. These electrons will produce a curvature radiation spectrum[13] with a characteristic energy given by

$$\varepsilon_c = \frac{3}{2}\hbar\frac{c}{R_c}\gamma^3. \qquad (5.3)$$

---

[13] Here we neglect possible contributions from synchrotron radiation.

While the particles radiate energy away, they are accelerated by an electric field $E_\parallel$ parallel to the magnetic field line, which we will express as a fraction η of the magnetic field $E_\parallel = \eta B$ (Gaussian units) assuming that the particles move close to the speed of light c. Equating the energy loss rate due to curvature radiation (left side) to the energy gain rate due to the electric field (right side),

$$\frac{2}{3}e^2 c \frac{\gamma^4}{R_c^2} = ec\eta B, \tag{5.4}$$

we obtain an upper limit for the Lorentz factor and can write the characteristic energy of our curvature spectrum as

$$\varepsilon_c = \left(\frac{3}{2}\right)^{7/4} \hbar c \left(\frac{\eta B}{e}\right)^{3/4} \sqrt{R_c}. \tag{5.5}$$

In addition, we assume a dipole-like magnetic field structure and replace B by its radial component $B = B_0 (R_0/R)^3$. Expressing the distances R and R_c in units of the light cylinder (LC) radius $R_{lc} = cP/2\pi$, that is $R = \xi R_{lc}$ and $R_c = \xi_c R_{lc}$, we get

$$\varepsilon_c = 150\,\text{GeV}\ \eta^{3/4}\ \xi^{-9/4}\sqrt{\xi_c}. \tag{5.6}$$

Even in the extreme case where the magnetic-field-aligned electric field approaches the strength of the magnetic field, which is at least a magnitude of order larger than predicted, and placing the emission site at $\xi = 2/3$, which is the inner boundary of the outer gap for an aligned rotator, we would need a curvature radius of $\xi_c \gtrsim 7$ to produce TeV photons, whereas current outer gap models usually consider the range $[0.3, 2]$ for ξ_c (see [28, 29], and references therein). We note that in recent dissipative pulsar magnetosphere models, curvature radiation from the current sheet or in its vicinity takes place with radii up to $\sim100\,R_{lc}$. But the corresponding electric field is not able to accelerate the particles to Lorentz factors beyond 10^8 and the resulting cut-off energies fall in the ~100 GeV range [30, 31], and references therein). Therefore, it seems highly unlikely that pulsed TeV photons from the Crab pulsar are produced via synchro-curvature radiation. A more plausible mechanism is inverse Compton (IC) scattering, which can occur for example in a synchrotron-self-Compton (SSC) model near the light cylinder as discussed by References [27, 32, 33], or in a wind zone model a few tens of light cylinder radii away from the pulsar as proposed by Aharonian et al. [34]. In any case, if we assume Compton scattering in the Klein-Nishina limit, the Lorentz factors of the accelerated electrons have to be greater than $\sim2\times10^6$ to produce 1 TeV photons. Although curvature radiation could account for the production of up to several hundred GeV photons, the simple power-law functions obtained by a joint fit of *Fermi*-LAT and MAGIC data from ~10 to ~1.2 TeV suggests the same mechanism to be at work above a few tens of GeV.

Theoretical interpretation

Concerning IC scattering we will consider two scenarios in greater detail that were proposed previously to explain pulsed emission up to ∼400 GeV from the Crab pulsar: the magnetospheric SSC model as discussed by Aleksić et al. [4] and IC scattering in a wind zone as proposed by Aharonian et al. [34]. Both scenarios are depicted in Fig. 5.10.

In the framework of the former, electron-positron pairs ($e^{\pm}$) are created and accelerated in an outer gap within the light cylinder, where they obtain large Lorentz factors of $\gamma \simeq 10^7$ producing curvature radiation in the 100–30 GeV range [32]. These primary $e^{\pm}$ also emit $\lesssim 5$ TeV gamma rays via IC scattering of magnetospheric infrared (IR) photons[14] that, in turn, are efficiently absorbed by the same IR field to materialize as secondary $e^{\pm}$ pairs with $\gamma \simeq 10^{4-7}$ (for a more detailed description of this cascading process, see Fig. 5.10). The lower-energy secondaries emit synchrotron photons in the eV to MeV range, while the higher-energy ones up-scatter such IR photons in the vicinity of the light cylinder to produce TeV gamma rays that reach the earth as the secondary SSC component of the pulsar. Similar ideas are also discussed in [27] with an analytical approach, and in Harding and Kalapotharakos [33] who assume that the primary $e^{\pm}$ are accelerated in a *slot gap* instead of an *outer gap*. Recently [35] also suggested that the $e^{\pm}$ could be accelerated not by electric fields along the magnetic field lines, but by the magneto-centrifugal force to reach Lorentz factors of ∼2×10^7 near the light cylinder (LC). Such energetic particles could also produce TeV gamma rays via IC scattering of thermal photons from the neutron star.

The alternative wind zone scenario by Aharonian et al. [34] considers the possibility of an emission region in the pulsar wind beyond the LC. It is generally accepted that the pulsar wind is initially dominated by electromagnetic energy near the LC but kinetically dominated when reaching the termination shock. It is still unclear where exactly the conversion happens and to what speed the wind is accelerated, which is also known as the sigma problem [36]. Therefore, Aharonian et al. [34] postulate a narrow cylindrical zone of radius between R_0 and R_f LC radii where abrupt creation and acceleration of $e^{\pm}$ take place with an initial Lorentz factor γ_0 to a maximum Lorentz factor of γ_w following an acceleration profile $\gamma(R) = \gamma_0 + (\gamma_w - \gamma_0)\left((R - R_0)/(R_f - R_0)\right)^{\alpha}$. These relativistic particles scatter with pulsed synchrotron X-rays from the pulsar via the IC process to produce gamma rays in the GeV to TeV regime. Their best fit model to previous data predicts a gradual acceleration with $\alpha = 1$ in a radius of $R_0 = 20R_{lc}$ and $R_w = 50R_{lc}$ assuming an initial and maximum Lorentz factor of $\gamma_0 = 300$ and $\gamma_w = 10^6$, respectively. However, these parameters lead to a strong cut-off at ∼500 GeV in the spectrum which is disfavored by our results (see Fig. 4.7). In principle one could recover compatible predictions with pulsed TeV emission by assuming a higher maximum Lorentz factor γ_w, extending the postulated acceleration zone further and adjusting the acceleration profile (see Fig. SM1 in Aharonian et al. [34]).

---

[14]The target photons are emitted from the cooling neutron star surface, from the heated polar cap surface, and from radiation processes in the magnetosphere in which pairs are created.

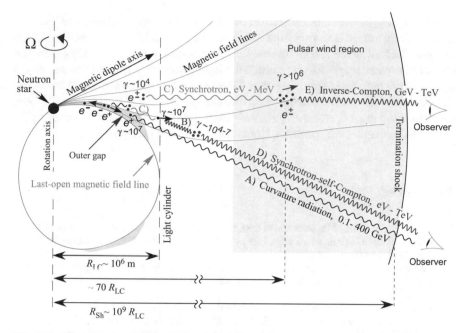

Fig. 5.10 Schematic figure of the Crab pulsar magnetosphere, the pulsar wind region (blue shaded region), and the synchrotron nebula at the termination shock. The neutron star, depicted by the black filled circle on the left, is rotating around the rotation axis (the vertical dashed line on the left). The thin black solid curves show the magnetic field lines. The pink region indicates the outer gap in which a strong electric field arises along the magnetic field lines as discussed in Aleksić et al. [4]. The black dots denote the primary electrons and positrons $e^{\pm}$ that are created and accelerated in the gap. These primary $e^{\pm}$ attain large Lorentz factors of $\gamma \simeq 10^7$ and efficiently emit 100–30 GeV photons (**A**) via the synchro-curvature (SC) process in the gap, and up to 5 TeV photons (**B**) via the inverse-Compton process of magnetospheric infrared (IR) photons after escaping from the gap. Some of the former photons collide with the magnetospheric X-ray photons (labeled as **C**) to materialize as secondary pairs (small red dots) with $\gamma \simeq 10^4$, while the remaining ones escape from the magnetosphere to be detected as pulsed signals mainly below 10 GeV. Since there could be an extreme acceleration due to the magneto-centrifugal force (see text), we denote the maximum attainable energy of these SC photons to be 400 GeV (see the long black wavy line towards the bottom right corner). The latter photons (**B**) are, in turn, absorbed by the magnetospheric IR photons (**C**) to materialize as secondary pairs (small red dots) with $\gamma \simeq 10^{4-7}$. The lower-energy secondaries ($\gamma \simeq 10^4$) efficiently emit synchrotron photons (**C**) in the eV to MeV range, while the higher-energy ones ($\gamma \simeq 10^7$) up-scatter such IR photons, which reach us as the synchrotron-self-Compton component below several TeV (**D**). This gives a possible explanation of our measurement of the pulsed TeV photons that appear in phase with the GeV pulsed signal. A distinct scenario was proposed by Aharonian et al. [34], who considers the inverse-Compton scatterings of the synchrotron X-rays (**C**) by ultrarelativistic $e^{\pm}$ in the pulsar wind. It is hypothesized in this model that an abrupt creation and acceleration of $e^{\pm}$ take place within a narrow cylindrical zone of radius of tens of light cylinder radii. Extending this scenario, we could in principle obtain TeV pulsed photons (**E**) by assuming greater maximum Lorentz factors $\gamma > 10^6$ than originally hypothesized (see text). For the sake of clarity, the photons of the first scenario (**A, D**) and the ones from the second (**E**) are depicted to propagate in different directions. They can, however, propagate into the same direction. Image courtesy of Kouichi Hirotani

In summary, several models seem to be able to explain qualitatively and partially quantitatively pulsed emission from the Crab pulsar up to 1 TeV. To reach such high energies they all make use of inverse Compton scattering of eV to MeV photons, either in the outer magnetosphere inside the light cylinder, or in a wind zone, tens of light cylinder radii away from the pulsar. Open issues that have yet to be addressed by theoretical models are the narrow peaks observed in the pulse profile above 100 GeV, the decreasing intensity ratio $P1/P2$ with energy and the phase coherence along the entire electromagnetical spectrum, from radio up to TeV energies.

References

1. Aleksić J et al (2015) Measurement of the Crab Nebula spectrum over three decades in energy with the MAGIC telescopes. J High Energy Astrophys 5–6:30–38. https://doi.org/10.1016/j.jheap.2015.01.002
2. Garrido Terrats D (2015) Limits to the violation of Lorentz invariance using the emission of the Crab pulsar at TeV energies, discovered with archival data from the MAGIC telescopes. PhD thesis
3. Ansoldi S et al (2016) Teraelectronvolt pulsed emission from the Crab Pulsar detected by MAGIC. Astron Astrophys 585:A133. https://doi.org/10.1051/0004-6361/201526853
4. Aleksić J et al (2012) Phase-resolved energy spectra of the Crab pulsar in the range of 50–400 GeV measured with the MAGIC telescopes. Astron Astrophys 540:A69. https://doi.org/10.1051/0004-6361/201118166
5. Zanin R et al (2013) MARS, the MAGIC analysis and reconstruction software. In: 33rd international cosmic ray conference, page id 773. Rio de Janeiro, Brazil
6. Hillas AM (1985) Cherenkov light images of EAS produced by primary gamma rays and by nuclei. In: 19th International cosmic ray conference, p 445. La Jolla, USA
7. Aleksić J et al (2012) Performance of the MAGIC stereo system obtained with Crab Nebula data. Astropart Phys 35(7):435–448. https://doi.org/10.1016/j.astropartphys.2011.11.007
8. Hobbs GB et al (2006) TEMPO2, a new pulsar-timing package - I. An overview. Mon Not R Astron Soc 369(2):655–672. https://doi.org/10.1111/j.1365-2966.2006.10302.x
9. Giavitto G (2013) Observing the VHE Gamma-Ray Sky with MAGIC Telescopes: the Blazar B3 2247+381 and the Crab Pulsar. PhD thesis
10. Lyne AG et al (1993) 23 years of Crab pulsar rotational history. Mon Not R Astron Soc 265(4):1003–1012. https://doi.org/10.1093/mnras/265.4.1003
11. Li TP, Ma YQ (1983) Analysis methods for results in gamma-ray astronomy. Astrophys J 272:317–324
12. Aliu E et al (2011) Detection of pulsed gamma rays above 100 GeV from the crab pulsar. Science 334(6052):69–72. https://doi.org/10.1126/science.1208192
13. Fierro JM et al (1998) Phase-resolved studies of the high-energy gamma-ray emission from the Crab, Geminga, and Vela pulsars. Astrophys J 494(2):734–746. https://doi.org/10.1086/305219
14. Aleksić J et al (2014) Detection of bridge emission above 50 GeV from the Crab pulsar with the MAGIC telescopes. Astron Astrophys 565:L12. https://doi.org/10.1051/0004-6361/201423664
15. de Jager OC et al (1989) A powerful test for weak periodic signals with unknown light curve shape in sparse data. Astron Astrophys 221:180–190
16. Albert J et al (2007) Unfolding of differential energy spectra in the MAGIC experiment. Nucl Instrum Methods Phys Res Sect A 583(2–3):494–506. https://doi.org/10.1016/j.nima.2007.09.048

17. Rolke WA et al (2005) Limits and confidence intervals in the presence of nuisance parameters. Nucl Instrum Methods Phys Res Sect A 551(2–3):493–503. https://doi.org/10.1016/j.nima.2005.05.068
18. Albert J et al (2008) VHE γ-ray observation of the crab nebula and its pulsar with the MAGIC telescope. Astrophys J 674(2):1037–1055. https://doi.org/10.1086/525270
19. Aleksić J et al (2016) The major upgrade of the MAGIC telescopes, part II: A performance study using observations of the Crab Nebula. Astropart Phys 72:76–94. https://doi.org/10.1016/j.astropartphys.2015.02.005
20. Eikenberry SS et al (1997) High time resolution infrared observations of the Crab Nebula pulsar and the pulsar emission mechanism. Astrophys J 477(1):465–474. https://doi.org/10.1086/303701
21. Abdo AA et al (2010) Fermi large area telescope observations of the Crab Pulsar and Nebula. Astrophys J 708(2):1254–1267. https://doi.org/10.1088/0004-637X/708/2/1254
22. Baring MG (2004) High-energy emission from pulsars: the polar cap scenario. Adv Space Res 33(4):552–560. https://doi.org/10.1016/j.asr.2003.08.020
23. Baring MG, Harding AK (2001) Photon splitting and pair creation in highly magnetized pulsars. Astrophys J 547(2):929–948. https://doi.org/10.1086/318390
24. Lee KJ et al (2010) Low bounds for pulsar γ-ray radiation altitudes. Mon Not R Astron Soc 405(3):2103. https://doi.org/10.1111/j.1365-2966.2010.16600.x
25. Muslimov AG, Harding AK (2003) Extended acceleration in slot gaps and pulsar high-energy emission. Astrophys J 588(1):430–440. https://doi.org/10.1086/368162
26. Du YJ et al (2012) Radio-to-TeV phase-resolved emission from the crab pulsar: the annular gap model. Astrophys J 748(2):84. https://doi.org/10.1088/0004-637X/748/2/84
27. Lyutikov M et al (2012) The very-high energy emission from pulsars: a case for inverse compton scattering. Astrophys J 754(1):33. https://doi.org/10.1088/0004-637X/754/1/33
28. Vigano D et al (2015) An assessment of the pulsar outer gap model-II. Implications for the predicted -ray spectra. Mon Not R Astron Soc 447(3):2649–2657. https://doi.org/10.1093/mnras/stu2565
29. Vigano D et al (2015) An assessment of the pulsar outer gap model - I. Assumptions, uncertainties, and implications on the gap size and the accelerating field. Mon Not R Astron Soc 447(3):2631–2648. https://doi.org/10.1093/mnras/stu2564
30. Grenier IA, Harding AK (2015) Gamma-ray pulsars: a gold mine. C R Phys 16(6–7):641–660. https://doi.org/10.1016/j.crhy.2015.08.013
31. Kalapotharakos C et al (2014) Gamma-ray emission in dissipative pulsar magnetospheres: from theory to fermi observations. Astrophys J 793(2):97. https://doi.org/10.1088/0004-637X/793/2/97
32. Hirotani K (2015) Three-dimensional non-vacuum pulsar outer-gap model: localized acceleration electric field in the higher altitudes. Astrophys J 798(2):L40. https://doi.org/10.1088/2041-8205/798/2/L40
33. Harding AK, Kalapotharakos C (2015) Synchrotron self-compton emission from the crab and other pulsars. Astrophys J 811(1):63. https://doi.org/10.1088/0004-637X/811/1/63
34. Aharonian FA et al (2012) Abrupt acceleration of a 'cold' ultrarelativistic wind from the Crab pulsar. https://doi.org/10.1038/nature10793
35. Osmanov Z, Rieger FM (2017) Pulsed VHE emission from the Crab Pulsar in the context of magnetocentrifugal particle acceleration. Mon Not R Astron Soc 464(2):1347–1352. https://doi.org/10.1093/mnras/stw2408
36. Arons J (2012) Pulsar wind nebulae as cosmic pevatrons: a current sheet's tale. Space Sci Rev 173(1–4):341–367. https://doi.org/10.1007/s11214-012-9885-1

Chapter 6
Lorentz Invariance Violation: Limits from the Crab Pulsar

Einstein has put an end to this isolation; it is now well established that gravitation affects not only matter, but also light.

Prof. H. A. Lorentz, 1920

The observation of pulsed photons up to $\sim$1 TeV provides a unique set of data to investigate fundamental physics. An application for which well timed emission at the highest energies is greatly valuable is testing for Lorentz invariance. Being one of the fundamental symmetries in relativity, any discovery of Lorentz Invariance Violation (LIV) would be an important signal of beyond the standard model physics [1]. The goal of the study presented in this chapter is to derive limits to the invariant energy scale, below which Lorentz violating effects, in terms of a wavelength dependent speed of light, are negligible. The study conducted by the author of this thesis was initially part of the cross-check procedure for the publication by Ahnen et al. [2]. In the following we present a substantially modified work based on the cross-check and use the extended Crab pulsar data set from the previous chapter.

In the first section we will give a short introduction to Lorentz invariance violation focusing on the experimental aspect of how to test it. The second section will explain in detail the statistical method we adopt to derive the limits. We conclude the chapter by reporting the results and discuss possible caveats, shortcomings and systematic uncertainties of our method.

6.1 A Short Introduction to Lorentz Invariance Violation

Although the violation of Lorentz symmetry is foreseen and incorporated in a number of theoretical ideas, quantum gravity (QG) models are arguably the most investigated class of theories that induce Lorentz invariance violation ([1], and references therein).

© Springer Nature Switzerland AG 2019
D. Carreto Fidalgo, *Revealing the Most Energetic Light from Pulsars and Their Nebulae*, Springer Theses, https://doi.org/10.1007/978-3-030-24194-0_6

The notion why a quantum theory of gravity should require a reassessment of Lorentz symmetry can be quickly motivated by the following line of argument. A common and model-independent feature of quantum gravity seems to be a minimum length or a lower bound to any output of a position measurement, presumably at the Planck scale [3]. If this minimum length would be measured to be l_p in the frame of one observer, it should be contracted by an amount $\sqrt{1 - v^2/c^2}$ for another observer traveling at the relative velocity v. This contradiction suggests that Lorentz invariance should break at very small distances [4]. A possible kinematic framework to describe this break is to propose a modified energy dispersion relation of the form $m^2 c^4 + p^2 c^2 = E\left[1 + f\,(E)\right]$, where m and p are the mass and momentum of the particle, respectively, c denotes the speed of light in vacuum and $f\,(E)$ is a QG model dependent function of the energy. Besides being a straight forward kinematic framework for Lorentz invariance violation, a modified dispersion relation has also the advantage that it can be embedded in an effective field theory as shown by Colladay and Kostelecký [5] in their minimal standard model extension.

In the following equations we will introduce a QG energy scale E_{QG}, which is generally believed to be of the order of the Planck scale ($E_p \approx 1.22 \times 10^{19}$ GeV). For small energies, that is for $E \ll E_{QG}$, we expect $f\,(E)$ to be well approximated by following series expansion and can write the dispersion relation for photons ($m = 0$) as (see [6])

$$p^2 c^2 = E^2 \left[1 + \sum_{n=1} \xi_n \left(\frac{E}{E_{QG_n}}\right)^n\right], \qquad (6.1)$$

where $\xi_n = \pm 1$ is a sign ambiguity that would be fixed in a given dynamical framework and corresponds to a *subluminal* scenario ($\xi_n = +1$, decreasing photon speed with increasing energy) and a *superluminal* case ($\xi_n = -1$, increasing photon speed with increasing energy), which will become clear from the next equation. Using the two Taylor series expansions $(1 + x)^{1/2} \approx 1 + x/2$ and $(x + 1)^{-1} \approx 1 - x$, where we identify x with the sum in Eq. 6.1, we can approximate the group velocity of photons with

$$v(E) = \frac{\partial E}{\partial p} \approx c \left[1 - \sum_{n=1} \xi_n \frac{n+1}{2} \left(\frac{E}{E_{QG_n}}\right)^n\right]. \qquad (6.2)$$

Meaningful constraints on the parameters E_{QG_n} in Eq. 6.2 can be obtained by time of flight measurements of photons from distant sources at different energies. The large distance to the source is needed to serve as an amplifier for the possible tiny differences in the photon speed that we would expect for photon energies much smaller than E_{QG_n}. Following this idea [6] were the first to propose the fast varying gamma-ray signals from astrophysical sources as test benches for Lorentz invariance. Since then various gamma-ray instruments in space and on the ground used observations of rapid flares in AGNs or GRBs to set lower limits on the energy scales E_{QG_n} (see, for example, [7–9]). A fast varying signal is needed to constrain the difference in arrival times Δt of photons at different energies, and hence to be more sensitive to a smaller difference in photon speeds Δv ($\Delta v \propto \Delta t/d$, d being the distance to the source). It is worth noting that constraints from AGN flares and GRBs are in a way complementary since they test different energy and distance ranges. On the one hand, very-high energy AGN flares have the advantage of being detected up to TeV energies with Cherenkov telescopes from the ground, but are limited in redshifts due to the absorption by the Extragalactic Background Light (EBL, [10]). Gamma-ray bursts on the other hand, are observed by satellites at large redshifts of up to $z \approx 8$, but are hardly detected above a few tens of GeV owing to the lack of photon statistics. Up to now the larger distances and shorter time scales of GRBs compared to AGN flares (seconds versus hundred of seconds) outweigh their lower energy range, and make limits derived from GRB observations the most stringent to date (see Table 6.1).

Another astrophysical source of varying gamma-ray signals are gamma-ray pulsars. Although they are observed many orders of magnitude closer than AGNs or GRBs, they exhibit much shorter time scales since their periodic pulses can be timed down to the order of microseconds. In contrast to AGN flares and GRBs, they also have the advantage of being persistent sources that can be observed on a regular basis to gather large amount of statistics and to systematically improve the derived limits. Moreover, the gradual slow-down of the pulsar's rotation theoretically allows to disentangle the ambiguity between energy dependent time delays intrinsic to the source and time delays induced by Lorentz invariance violation. While intrinsic time delays would be observed proportional to the rotational phase of the pulsar, LIV-induced time delays would be unaffected by the slow-down and would show up as decreasing phase shifts in the pulsar's light curve with time [12].

The first limits on LIV energy scales using a gamma-ray pulsar, namely the Crab pulsar, were derived by Kaaret [13] who analyzed data up to 2 GeV from the Energetic Gamma-Ray Experiment Telescope (EGRET, [14]). These limits were improved by Otte [12] taking advantage of the very-high-energy tail of the Crab pulsar discovered in 2011 by the Cherenkov telescopes VERITAS and MAGIC (see Chap. 4). With the extension of this tail up to ~1 TeV energies, as reported in the previous chapter of this thesis, we will try to use this work's unique data set to challenge the current

Table 6.1 A selection of recent limits on the parameters $E_{QG_{1,2}}$ (subluminal case, see text) obtained from AGN flares and GRB observations by means of time of flight measurements

References	Instrument	Source	95% CL lower limits
Martínez and Errando [11]	MAGIC	Mrk 501	$E_{QG_1} > 3.0 \times 10^{17}$ GeV
			$E_{QG_2} > 5.7 \times 10^{10}$ GeV
Abramowski et al. [7]	H.E.S.S.	PKS 2155-304	$E_{QG_1} > 2.1 \times 10^{18}$ GeV
			$E_{QG_2} > 6.4 \times 10^{10}$ GeV
Vasileiou et al. [9]	*Fermi*-LAT	GRB090510	$E_{QG_1} > 9.3 \times 10^{19}$ GeV
			$E_{QG_2} > 1.3 \times 10^{11}$ GeV

Notes If references obtained results via various statistical methods, we state their best limits. Higher order terms ($n > 2$) are usually ignored since the sensitivity of current experiments is not sufficient to deduce meaningful limits on them, that is many orders of magnitude away from the Planck scale

world-best limits by Vasileiou et al. [9], stated in Table 6.1. Especially the constraint on the second order term E_{QG_2} should profit from the higher energy scale compared to GRBs. That said, it is interesting to note that in the framework of the minimal standard model extension, [5] derive a violation of CPT for odd terms in n and showed that the operators corresponding to a linear variation of the speed of light also rotate the linear polarization of photons quadratic in energy. The observation of this predicted birefringence in the optical, recently allowed [15] to place a stringent lower limit of 10^6 times the Planck energy on the linear term E_{QG_1}. Against this background, the quadratic term E_{QG_2} seems currently to be of particular interest. For higher order terms the sensitivity of present instruments only allows to deduce limits many orders of magnitude away from the Planck scale. Therefore, terms with $n > 2$ are usually ignored in modern LIV searches.

For an extensive discussion about the whole range of experimental LIV tests, including astrophysical as well as terrestrial constraints, the reader is referred to the excellent review by Mattingly [1] and references therein.

6.2 Our Statistical Method

In the case of pulsars searching for an energy dependence of the photon speed basically means to look for phase shifts of the pulse with energy. This shift $\Delta\phi$ between a low and high energy photon, E_l and E_h, can be expressed by means of Eq. 6.2 as (considering only the n-th term of the sum)

$$\Delta\phi = \phi_h - \phi_l = \frac{d\ (v_l - v_h)}{v_h\,v_l\,P} = \frac{d}{c\,P}\,\xi_n\,\frac{n+1}{2}\,\frac{E_h^n - E_l^n}{E_{QG_n}^n}, \tag{6.3}$$

where d and P are the distance and the period of the pulsar, respectively, and $v_{l,h}$ are the group velocities of the low and high energy photon. Here we also assume that the sum of Eq. 6.2 is much smaller than 1 for the photon energies considered. The simplest approach to constrain E_{QG_n} would now be to compare the peak position in the pulsar's phaseogram for two energy bands and derive limits from $E_{QG_n} \gtrsim \Delta\phi^{1/n}$. A more sensitive approach, which takes into account single events and exploits the full information contained in the data set, is to construct a likelihood function for the measurement by modeling the emission of the pulsar, the background and the instrument response. This approach was first advocated for the same kind of LIV searches by Martínez and Errando [11] using AGN flares and was also applied by Vasileiou et al. [9] for their GRB analyses.

Before describing our likelihood function and the corresponding model in detail we will formally introduce two new parameters that we refer to as the linear and the quadratic LIV parameter: $E_1^* \equiv E_{QG_1}[\text{GeV}]/10^{19}$ and $E_2^* \equiv E_{QG_2}[\text{GeV}]/10^{10}$. These parameters are directly proportional to the LIV energy scales $E_{QG_{1,2}}$ and are chosen in a way that competitive limits on them should lie around unity. Hence, for photons with energy E coming from the Crab pulsar, a phase shift due to the LIV effect can then be written as

$$\Delta\phi\left(E, E_n^*\right) = \xi_n \frac{d_{\text{Crab}}}{c\, P_{\text{Crab}}} \frac{n+1}{2} 10^{-18-n} \left(E/E_n^*\right)^n \quad \text{for} \quad n \in \{1, 2\}. \quad (6.4)$$

The distance to the Crab pulsar d_{Crab} is still determined rather poorly and we adopt the generally stated value of 2 ± 0.5 kpc [16, 17]. For its period P_{Crab} we average over all the ephemerides used in the analysis of Chap. 5 and obtain a value of $33.65^{+0.04}_{-0.05}$ ms, where the errors depict the maximum and minimum value found in the ephemerides. This error is negligible compared to the error of the distance estimate.

For our likelihood we will only consider events around the inter-pulse of the Crab pulsar's phaseogram. We refrain from taking advantage of the full information from the whole phase range, since background events do not carry any information about the LIV effect and photons from the main pulse should be far less constraining on the LIV energy scale due to their limited energy range (see previous chapter). In addition, modeling the main pulse and Crab's bridge emission would needlessly complicate the likelihood and introduce further systematic uncertainties. Also from a computational point of view, it is much more efficient to limit the phase range as much as possible. In the following we will refer to the inter-pulse phase region simply as *On region*. We limit this On region to $\phi_{on} = [0.35, 0.45]$ in phase, which is motivated by two estimations: (i) extrapolating the power-law found in the previous chapter for the inter-pulse, we expect on average ~ 1 event above 10 TeV from the inter-pulse; (ii) assuming the best LIV limits stated in Table 6.1, we would expect a phase shift of roughly 0.05 at 10 TeV (see Fig. 6.1). Thus, we choose ± 0.05 around the inter-pulse at ~ 0.4 in phase.

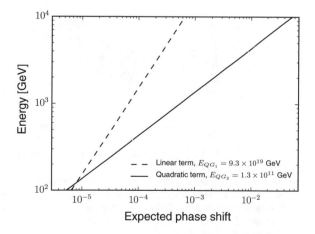

Fig. 6.1 Theoretical phase shifts in the Crab pulsar following Eq. 6.4 and assuming the current world-best limits (see Table 6.1). At 10 TeV we would expect a phase shift of ~ 0.05 due to the quadratic term, and hence should obtain more competitive limits for the quadratic than for the linear term

Linear term, $E_{QG_1} = 9.3 \times 10^{19}$ GeV
Quadratic term, $E_{QG_2} = 1.3 \times 10^{11}$ GeV

With the help of the following model we will try to find probability density functions (PDFs) for the events in the On region of the phaseogram, and subsequently construct our likelihood function. Following the results of the previous chapter, we assume a simple power law for the energy distribution $\Gamma(E)$ of the photons coming from the Crab pulsar and approximate their distribution in phase $\Psi(\phi, E)$ by a Gaussian, in which we incorporate the phase shift due to a possible LIV effect:

$$\Gamma\left(E \mid f_{E_0}, \alpha\right) = f_{E_0}(E/E_0)^{-\alpha} \tag{6.5}$$

$$\Psi\left(\phi, E \mid \mu_{P2}, \sigma_{P2}; E_n^*\right) = \frac{1}{\sqrt{2\pi}\,\sigma_{P2}} \exp\left[-\frac{\left(\phi - \mu_{P2} - \Delta\phi\left(E \mid E_n^*\right)\right)^2}{2\sigma_{P2}^2}\right]. \tag{6.6}$$

If $(\phi - \Delta\phi) \notin [0, 1)$, we project the resulting phase back to $[0, 1)$ before evaluating Eq. 6.6.

The instrument response is characterized by two quantities: the collection area $C(E)$ and the migration matrix $M(E, E')$. While the collection area represents the effective size of a perfect detector, the migration matrix takes into account the finite energy resolution of the instrument (see paragraph *Flux and Spectrum calculation* in Sect. 3.4 for details). Since the instrument response is usually binned coarsely in energy to ensure sufficient MC statistics in each bin, we upsample both quantities to approximate a nearly unbinned response. Figure 6.2 shows the instrument response $R(E, E')$ defined as the product of the migration matrix and the collection area.

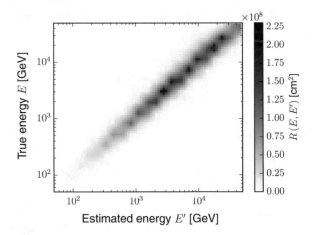

Fig. 6.2 Our instrument response defined as the product of the migration matrix and the collection area, $R = M \times C$. By interpolating linearly we upsample the response to a 100×100 grid (see text). Some insignificant artifacts of the original binning are still visible. The collection area $C(E)$ and the migration matrix $M(E, E')$ are taken from the analysis of Sect. 5.5 and are provided by the MARS executable *CombUnfold* (see Sect. 3.4)

The source model together with the instrument response allows us to specify a PDF for the signal, that is events coming from the inter-pulse of the Crab pulsar:

$$S\left(E', \phi \mid \nu'\right) = A \int_E R(E, E') \; \Gamma\left(E \mid f_{E_0}, \alpha\right) \; \Psi\left(\phi, E \mid \mu_{P2}, \sigma_{P2}; E_n^*\right) \; dE, \quad (6.7)$$

where ν' contains our model parameters $\{E_n^*, f_{E_0}, \alpha, \mu_{P2}, \sigma_{P2}\}$ and A is the normalization constant that ensures $\int_{\phi_{on}} \int_{E'} S \, dE' d\phi = 1$.

The estimated energy distribution of background events in the On region is a result from a rather complex combination of cosmic-ray events, the flux from the Crab nebula and the corresponding instrument response, and hence is hard to model analytically. Instead we choose to linearly interpolate a binned histogram of estimated energy in double logarithmic space, making use of the events falling in the Off phase region [0.52, 0.87] (see previous chapter). The choice of binning is a compromise between including enough statistics in each bin and catching at the same time the subtle features present in the distribution (see Fig. 6.3). It is important to incorporate those features and model the distribution accurately, since the integrated number of background events is always an order of magnitude larger than the number of signal events. The normalized background distribution, that is $\int_{\phi_{on}} \int_{E'} h \, dE' d\phi = 1$, gives us the background PDF $h(E')$.

The limited capability of modeling the background also influences the range in estimated energy that we will consider for our likelihood. The small rise of the distribution above $\sim$100 GeV is likely caused by the onset of the events from the mono data (see Sect. 5.3), while below $\sim$100 GeV the shape of the distribution is more complex and becomes harder to model accurately. Another argument for discarding low energy events in favor of a more efficient computing time, is the fact that we primarily expect the high energy events from the Crab pulsar (> 400 GeV) to drive the constraining power on the LIV energy scales. Above $\sim$200 GeV the distribution seems to follow roughly a power law. Above $\sim$10 TeV the limited statistics start

Fig. 6.3 The probability density function for the background events. We derive the PDF by linearly interpolating a histogram with 150 bins in estimated energy in double logarithmic space. The histogram is filled with events falling in the Off phase region [0.52, 0.87]

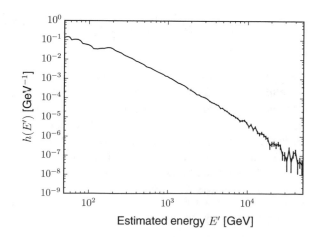

to dominate the shape and we do not expect more than ~ 1 event on average above 10 TeV, as discussed before. Therefore, we choose to include only events with an estimated energy between 100 GeV and 10 TeV for our likelihood. In addition, the lower limit of 100 GeV also facilitates a comparison to the results obtained in the previous chapter, in which the minimum energy for the light curve study was set to 100 GeV as well.

The signal and background PDFs, S and h, respectively, now allow us to construct a PDF for the events in the On region:

$$P\left(E', \phi \mid \nu\right) = \frac{g(\nu')S\left(E', \phi \mid \nu'\right) + b\, h\left(E'\right)}{g\left(\nu'\right) + b}, \tag{6.8}$$

where:

- $g\left(\nu'\right) = t_{\text{eff}} \times \int_{\phi_{on}} \int_{E'} S/A\, dE' d\phi$ is the total number of signal events in our On region for the effective observation time t_{eff},
- b is the total number of background events in our On region and is treated as a free model parameter,
- ν contains all our free model parameters $\{E_n^*, f_{E_0}, \alpha, \mu_{P2}, \sigma_{P2}, b\}$,
- and the denominator $g\left(\nu'\right) + b$ serves as the normalization constant.

Alongside P, in our likelihood we simultaneously fit the total number of events in the Off phase region N_{off} with a Poisson distribution to guide the estimation of b:

$$B(N_{\text{off}} \mid b) = e^{-\tau b} \frac{(\tau b)^{N_{\text{off}}}}{N_{\text{off}}!}, \tag{6.9}$$

where τ is the ratio between the phase widths of the background and On region. Since we expect τb to be of the order of 10^5, we can safely approximate Eq. 6.9 by a Gaussian around τb with a standard deviation of $\sqrt{\tau b}$, which is computationally more convenient. Our likelihood function then reads:

$$\mathcal{L}\left(\nu \mid \{E_i', \phi_i\}_{i=0}^{N_{\text{on}}}, N_{\text{off}}\right) = B\left(N_{\text{off}} \mid b\right) \prod_{i=0}^{N_{\text{on}}} P\left(E_i', \phi_i \mid \nu\right), \tag{6.10}$$

where the product $\prod_{i=0}^{N_{\text{on}}}$ runs over all events in the On region.

With the likelihood function at hand, we will resort to Bayesian inference to estimate our model parameters and derive limits on the LIV energy scales [20]. This choice is partially motivated by the attempt to avoid the cumbersome calibration of the test statistics in the case of a maximum likelihood approach, as extensively carried out by Ahnen et al. [2]. One draw back of the Bayesian inference approach is arguably the need for defining priors. Prior information about the parameters can be found in essentially 3 publications that deal with the very-high-energy tail of the Crab pulsar above 100 GeV (see Chap. 4). While results by Aliu et al. [18] were obtained from completely independent observations by VERITAS, the MAGIC papers

[19, 21] partly use a smaller and bigger subset, respectively, of the data considered in this work. Therefore, we prefer the former findings to define our priors for the pulse shape, that is μ_{P2} and σ_{P2}. For the spectral parameters, f_{E_0} and α, we have to fall back to the inter-pulse spectrum of [19], since [18] only report numbers for the phase-averaged case including the Crab pulsar's main pulse. For all our model parameters we choose rather weak priors in the sense that we assume normal distributions with a sigma of 3 times the statistical errors adding in quadrature the systematical error, if stated. With this choice we aim to take into account the partially different energy ranges considered in [18, 19], compared to our data set.

For our LIV parameter E_n^* we choose a log-uniform prior $(1/x)$, which is a common choice when seeking a non-informative prior for a scale parameter. However, since this is an improper prior and we do not expect our likelihood to go to zero for arbitrarily high E_n^*, we have to artificially introduce an upper limit to assure a proper posterior. A conservative approach is to set this limit to reflect our systematic uncertainty in the pulsar phases of the events. At 10 TeV, which is the highest energy considered in our analysis, a LIV energy scale of $E_{QG_1} \gtrsim 2 \times 10^{19}$ and $E_{QG_2} \gtrsim 5 \times 10^{11}$, respectively, would lead to a phase shift smaller than our systematics, that is $\Delta\phi \lesssim 0.0039$. We do not expect our analysis to limit the scales beyond these values, and therefore take them as upper limits for the priors. Table 6.2 summarizes all priors for our model parameters.

Table 6.2 Priors for our model parameters

Parameter	Prior	Assumption
b	Uniform: $\begin{cases} 1, & b \in [0, 10^{10}] \\ 0, & b \notin [0, 10^{10}] \end{cases}$	Non-informative
μ_{P2}	Normal: $\exp\left[-\frac{1}{2} \frac{(0.3978 - \mu_{P2})^2}{0.006^2} \right]$	Aliu et al. [18]
σ_{P2}	Normal: $\exp\left[-\frac{1}{2} \frac{(0.0113 - \sigma_{P2})^2}{0.066^2} \right]$	Aliu et al. [18]
f_{E_0}	Normal: $\exp\left[-\frac{1}{2} \frac{(4.7 - f_{E_0})^2}{2.5^2} \right]$	Aleksić et al. [19]
α	Normal: $\exp\left[-\frac{1}{2} \frac{(-3.42 - \alpha)^2}{0.81^2} \right]$	Aleksić et al. [19]
E_1^*	Log-uniform: $\begin{cases} 1/E_1^*, & E_1^* \in [10^{-19}, 2] \\ 0, & E_1^* \notin [10^{-19}, 2] \end{cases}$	Non-informative
E_2^*	Log-uniform: $\begin{cases} 1/E_2^*, & E_2^* \in [10^{-10}, 50] \\ 0, & E_2^* \notin [10^{-10}, 50] \end{cases}$	Non-informative

Notes The assumption, which the prior is based on, is specified in the third column (see text). The background parameter b is well constrained by our likelihood function and reasonable choices for its limits do not affect the outcome. We also try to use a log-uniform prior but do not find any significant differences in the results. The same holds for the lower limits of our LIV parameters and we choose values corresponding to an energy scale of $E_{QG_n} = 1$ GeV. The parameter f_{E_0} is given in units of 10^{-11} cm^{-2} s^{-1} TeV^{-1} at $E_0 = 120$ GeV

Having defined our likelihood function and our priors, we draw samples from the posterior distributions by employing the Markov chain Monte Carlo (MCMC) Ensemble sampler *emcee*[1] [22]. We run the sampler with 200 walkers, discard the first 300 steps due to a *burn-in* phase and save the subsequent 500 steps to obtain a representative sample of the posterior distributions. A successful burn-in is verified by visually inspecting the MCMC chains and by checking for a healthy acceptance rate of the proposed states. Furthermore, we calculate autocorrelation times to assure the representativeness of our samples. We state our uncertainties for the model parameters as 16th, 50th and 84th percentiles of the samples in the obtained distributions.

6.3   Results and Discussion

This analysis uses the data set from the spectral calculations of Sect. 5.5. The main goal is to derive in total 4 lower limits, that is a subluminal and a superluminal one ($\xi_n = \pm 1$) for the linear and the quadratic LIV parameter, respectively, which we will refer to as E_1^{*+}, E_1^{*-}, E_2^{*+} and E_2^{*-}.

As a first check for our statistical approach we try to reproduce the results obtained in Chap. 5. For this we estimate our model parameters fixing the LIV parameters to the upper limits discussed in the previous section. These limits correspond to the smallest phase shifts detectable in our data set and can therefore be interpreted as a scenario where no LIV effect is present. The results of the MCMC run are displayed in Fig. 6.4, in which we show the posterior distributions of the parameters and their correlations with each other. A quick comparison to the previous estimates is given in Table 6.3. The values of the two analyses agree well within their statistical errors, except for the peak width. We attribute the slight discrepancy to the fact that the value stated from Chap. 5 is taken from the (100–400) GeV pulsar light curve, whereas this analysis incorporates events up to 10 TeV. And as mentioned in Sect. 5.4, we see a hint of peak broadening with energy. Although both analyses are based on the same data set, we want to point out a couple of relevant differences between them:

- For the estimation of the pulse shape in Sect. 5.4 we applied different signal extraction cuts that maximized the significance of the detection.
- For the spectral analysis in Sect. 5.5 we only considered the number of excess events in the estimated energy range from ~(80–1700) GeV in the narrow phase range of [0.377, 0.422], and made no assumption about the peak shape.

Table 6.3 also compiles the results from Ahnen et al. [2] that were obtained by means of a maximum likelihood method in the energy range ≥ 100 GeV. Considering the use of slightly different data sets as well as differences in the employed methods, which will be discussed in more detail at the end of the chapter, the results seem to be consistent.

---

[1] http://dan.iel.fm/emcee/current, last accessed 03/02/2018.

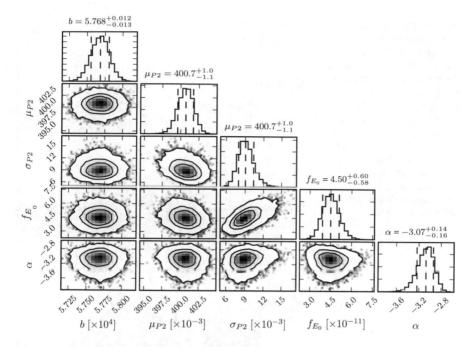

Fig. 6.4 Posterior distributions of the model parameters and their correlations with each other. For this MCMC simulation we assumed a quadratic dependence on the energy for a subluminal photon speed ($\xi = +1$) and fix its parameter to $E_2^* = 50$ (see previous section). Compatible distributions and correlations are obtained when assuming a superluminal scenario or a linear dependence and fixing its parameter to $E_1^* = 2$. The contours in the 2D plots show 1, 2 and 3 sigma deviations. The dashed lines in the histograms denote the 16th, 50th and 84th percentiles that are also mentioned in the corresponding figure titles. For the plotting we use the python package *corner* [23]

Table 6.3 Comparison of the model parameters with previous results from this thesis and the results from Ahnen et al. [2]

Parameter	Bayesian inference	Analysis from Chap. 5	Ahnen et al. [2]
μ_{P2} [$\times 10^{-3}$]	$400.7^{+1.0}_{-1.1}$	$401.4^{+1.4}_{-1.5}$	401 ± 1
σ_{P2} [$\times 10^{-3}$]	$9.0^{+1.3}_{-1.1}$	$7.8^{+1.6}_{-1.3}$	11 ± 2
f_{E_0} [$\times 10^{-11}$]	$4.50^{+0.60}_{-0.58}$	4.59 ± 0.40	4.7 ± 0.5
α	$-3.07^{+0.14}_{-0.16}$	3.13 ± 0.18	2.95 ± 0.07

Notes For the Bayesian inference method, compatible results are obtained when assuming a linear dependence on the energy or a superluminal scenario. The results from Ahnen et al. [2] were obtained by means of a maximum likelihood method in the energy range ≥ 100 GeV. f_{E_0} is given in units of [cm^{-1} s^{-1} TeV^{-1}] with $E_0 = 120$ TeV (see Sect. 5.5)

Fig. 6.5 Comparison between the model prediction and the data for the phaseogram in 3 different energy ranges. We assume a quadratic dependence on the energy for a subluminal photon speed ($\xi = +1$) and fix its parameter to $E_2^* = 50$ (see previous section). The distribution of the normalized residuals is shown in the figure's inset and should follow a Gaussian distribution (see text)

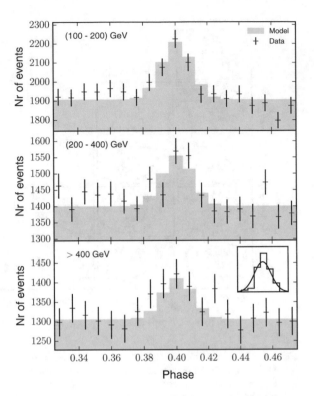

Another check to see if our model adequately describes the data, is to perform a test for the goodness of fit. For this we simulate binned phaseograms of our model in three energy ranges and compare the prediction to the data itself (see Fig. 6.5). The simulated histograms are easily obtained by integrating Eq. 6.8 leaving out the normalization constant. For the true model having the true parameter values, the normalized residuals (that is divided by the a priori known measurement errors) should follow a normal distribution with zero mean and a variance of one [24]. Hence, we check for normality of our distribution (see inset of Fig. 6.5) by applying a Kolmogorov-Smirnov and Shapiro-Wilk test, but find no significant deviation from a Gaussian.

After doing these basic checks, we now perform 4 MCMC runs in which we free the LIV parameters and assume a sub- and superluminal case for each one of the two. The resulting posterior distributions of the LIV parameters are shown in Fig. 6.6 together with the relations to the other model parameters. As expected the posterior distributions seem to reach a plateau towards the upper limits reflecting the insensitiveness of our data to even higher LIV parameters. Another intuitive result is the visible correlation with the peak position μ_{P2}, since its shift is able to mimic the LIV effect. The modes of the posterior distributions do not coincide with the upper limits and hint to an actual preference of a peak shift over no shift at all. In each case, however, the statistical significance is around or below 1 sigma, which we estimate

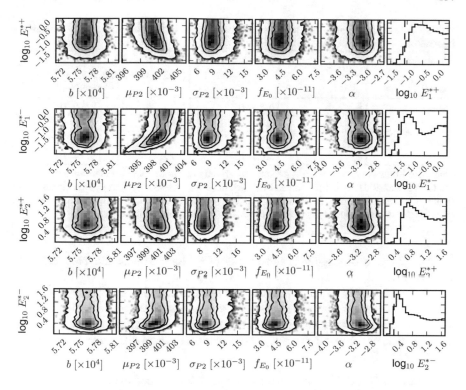

Fig. 6.6 The results from our 4 MCMC simulations (see text). The right column shows the posterior distributions of the linear and quadratic LIV parameter for a sub- and superluminal scenario. The dashed lines denote the 5th percentiles. The remaining panels show the correlations with the other model parameters and 1, 2 and 3 sigma deviations

by applying a likelihood ratio test between the mode value of the LIV parameter and its upper limit fixing the rest of the model parameters.

Due to the implementation of upper limits to $E_{1,2}^*$, the fifth percentiles of the posterior distributions translate to conservative 95% credible intervals for $E_{1,2}^* \in (E_{1,2}^{*5th}, \infty]$, meaning that the chance of the parameter lying in the interval is *at least* 95%. We transform these percentiles to lower limits on the LIV energy scales $E_{QG_{1,2}}$, including the systematic uncertainty of the Crab pulsar's distance,[2] and compare them in Table 6.4 to the 95% confidence intervals obtained by Ahnen et al. [2] and the current world-best limits by Vasileiou et al. [9]. Our limits differ from those derived by Ahnen et al. [2] up to a factor of $\sim$3.3. Even though both analysis are based on a similar data set, there are important differences between them, which we will mention further down. Compared to the world-best limits our results fall

---

[2]Following Eq. 6.3 this means a worsening of 25 and 13% for the linear and quadratic limit, respectively.

Table 6.4 Comparison of the 95% credible intervals derived in this work and the 95% confidence intervals from previous publications

Parameter [GeV]	Scenario	This work	Ahnen et al. [2] (incl. systematics)	[9] (best of 3 methods)
E_{QG_1}	$\xi = +1$	6.3×10^{17}	5.5×10^{17}	9.3×10^{19}
	$\xi = -1$	2.6×10^{17}	4.5×10^{17}	1.3×10^{20}
E_{QG_2}	$\xi = +1$	3.1×10^{10}	5.9×10^{10}	1.3×10^{11}
	$\xi = -1$	1.6×10^{10}	5.3×10^{10}	9.4×10^{10}

Notes Our limits only include the systematic uncertainty due to the poorly determined distance to the Crab pulsar of 2.0 ± 0.5 pc. Ahnen et al. [2], however, consider various systematic errors for their limits, which are discussed in the text. The limits quoted from [9] are the best out of 3 statistical methods and do not include systematic uncertainties due to instrumental effects

behind $\gtrsim 2$ orders of magnitude and a factor of $\lesssim 6$ for the linear and quadratic term, respectively.

We want to emphasize that the study presented here is far from complete and has several shortcomings. Being essentially a cross-check for the main results in Ahnen et al. [2], we simplified a couple of approaches and omitted the study of systematic uncertainties, which is out of scope for this work. The main differences compared to Ahnen et al. [2] are:

- While Ahnen et al. [2] used a maximum likelihood method to derive their approximated 95% confidence limits, we applied Bayesian inference to obtain conservative 95% credible intervals.
- Instead of differentiating between the different instrument responses corresponding to the analysis epochs and settings of the data (see Table 5.1), we took an average using the observation times as weights, which is the same approach as the one of the spectral analysis in Sect. 5.5.
- Instead of defining a fit function to achieve a truly unbinned response, we upsampled the coarse binning of the effective area and migration matrix.
- We do not propagate the statistical errors of the instrument response to the final model parameters.
- While Ahnen et al. [2] only considered events with a reconstructed energy of ≥ 400 GeV, we restrict our energy range from 100 GeV to 10 TeV.
- We did not check for a possible bias in the estimation of our LIV parameters. This can be done by producing mock data sets from our model assuming various finite LIV energy scales and feed them to our Bayesian inference analysis (which is rather time consuming computationally wise).
- The systematic uncertainty of the phase determination is a factor of a few smaller than the phase width of the peak, yet should be implemented properly in the likelihood function of the model.
- Important systematic uncertainties that we did not account for, but which were properly addressed in Ahnen et al. [2], include: the background estimation (that is the determination of $h(E')$), systematic errors in the instrument response, potential

cutoffs in the pulsar's spectrum, different pulse shapes and its evolution with energy, a possible contribution of the bridge emission to the excess.

Without properly addressing these differences, the comparison between the two analyses is hard to interpret. Regarding the different statistical treatments, a clear disadvantage of our method is the need for introducing an upper limit on the LIV parameter, and thus the construction of conservative intervals for the LIV energy scales. In this case a profile likelihood ratio method, as employed by Ahnen et al. [2] seems to be the more adequate approach, at the expense of a cumbersome calibration of the corresponding test statistic. We conclude, however, that by applying an independent statistical method and computational code to the related data sets and by obtaining limits of the same order of magnitude, we corroborate the principal findings by Ahnen et al. [2].

As mentioned in the the beginning of this chapter, limiting the quadratic term E_{QG_2} seems at present to be of greater interest given the strong constraints on the linear term. The quadratic limits derived in this work and by Ahnen et al. [2] are approximately an order of magnitude better than the ones presented by Otte [12] using the Crab pulsar, and are within reach of the current world-best limits obtained by means of GRB observations [9]. With the upcoming next generation Cherenkov Telescope Array (CTA, see Sect. 1.1) we expect a better resolution of the pulse shape and a better understanding of its evolution with energy, which in turn should enable us to improve our model and consequently our limits. In their study Ahnen et al. [2] include simulations with a toy performance of CTA and conclude that even with a factor ten less observation time dedicated to the Crab pulsar, their present limits will be improved. Gamma-ray pulsars can therefore be reintroduced to the class of astrophysical objects that allow useful and competitive studies of energy dependent photons speeds.

References

1. Mattingly D (2005) Modern tests of Lorentz invariance. Living Rev Relativ 8(1):5. https://doi.org/10.12942/lrr-2005-5
2. Ahnen ML et al (2017b) Constraining Lorentz invariance violation using the crab pulsar emission observed up to TeV energies by MAGIC. Astrophys J Suppl Ser 232(1):9. https://doi.org/10.3847/1538-4365/aa8404
3. Garay LJ (1995) Quantum gravity and minimum length. Int J Mod Phys A 10(02):145–165. https://doi.org/10.1142/S0217751X95000085
4. Plato ADK et al (2016) Gravitational effects in quantum mechanics. Contemp Phys 57(4):477–495. https://doi.org/10.1080/00107514.2016.1153290
5. Colladay D, Kostelecký VA (1998) Lorentz-violating extension of the standard model. Phys Rev D 58(11):116002. https://doi.org/10.1103/PhysRevD.58.116002
6. Amelino-Camelia G et al (1998) Tests of quantum gravity from observations of gamma-ray bursts. Nature 393(6687):763–765. https://doi.org/10.1038/31647
7. Abramowski A et al (2011) Search for Lorentz Invariance breaking with a likelihood fit of the PKS 2155–304 flare data taken on MJD 53944. Astropart Phys 34(9):738–747. https://doi.org/10.1016/j.astropartphys.2011.01.007

8. Albert J et al (2008a) Probing quantum gravity using photons from a flare of the active galactic nucleus Markarian 501 observed by the MAGIC telescope. Phys Lett B 668(4):253–257. https://doi.org/10.1016/j.physletb.2008.08.053

9. Vasileiou V et al (2013) Constraints on Lorentz invariance violation from fermi -large area telescope observations of gamma-ray bursts. Phys Rev D 87(12):122001. https://doi.org/10.1103/PhysRevD.87.122001

10. Mazin D et al (2013) Potential of EBL and cosmology studies with the Cherenkov telescope array. Astropart Phys 43:241–251. https://doi.org/10.1016/j.astropartphys.2012.09.002

11. Martínez M, Errando M (2009) A new approach to study energy-dependent arrival delays on photons from astrophysical sources. Astropart Phys 31(3):226–232. https://doi.org/10.1016/j.astropartphys.2009.01.005

12. Otte AN (2011) Prospects of performing Lorentz invariance tests with VHE emission from pulsars. In: 32nd international cosmic ray conference. Beijing, China. https://doi.org/10.7529/ICRC2011/V07/1302

13. Kaaret P (1999) Pulsar radiation and quantum gravity. Astron Astrophys 345:32–34

14. Thompson DJ (2008) Gamma ray astrophysics: the EGRET results. Rep Prog Phys 71(11):116901. https://doi.org/10.1088/0034-4885/71/11/116901

15. Kislat F, Krawczynski H (2017) Planck-scale constraints on anisotropic Lorentz and CPT invariance violations from optical polarization measurements. Phys Rev D 95(8):083013. https://doi.org/10.1103/PhysRevD.95.083013

16. Trimble V (1973) The distance to the crab nebula and NP 0532. Publ Astron Soc Pac 85(October):579. https://doi.org/10.1086/129507

17. Kaplan DL et al (2008) A precise proper motion for the crab pulsar, and the difficulty of testing spin-kick alignment for young neutron stars. Astrophys J 677(2):1201–1215. https://doi.org/10.1086/529026

18. Aliu E et al (2011) Detection of pulsed gamma rays above 100 GeV from the crab pulsar. Science 334(6052):69–72. https://doi.org/10.1126/science.1208192

19. Aleksić J et al (2012b) Phase-resolved energy spectra of the Crab pulsar in the range of 50–400 GeV measured with the MAGIC telescopes. Astron Astrophys 540:A69. https://doi.org/10.1051/0004-6361/201118166

20. MacKay DJ (2005) Information theory, inference, and learning algorithms. J Am Stat Assoc 100(472):1461–1462. https://doi.org/10.1198/jasa.2005.s54

21. Aleksić J et al (2014a) Detection of bridge emission above 50 GeV from the crab pulsar with the MAGIC telescopes. Astron Astrophys 565:L12. https://doi.org/10.1051/0004-6361/201423664

22. Foreman-Mackey D et al (2013) emcee : the MCMC Hammer. Publ Astron Soc Pac 125(925):306–312. https://doi.org/10.1086/670067

23. Foreman-Mackey D (2016) corner.py: Scatterplot matrices in Python. J Open Sour Softw 1(2). https://doi.org/10.21105/joss.00024

24. Andrae R et al (2010) Dos and don'ts of reduced chi-squared. ArXiv, ID 1012:3

Chapter 7
Search for the Next Very-High-Energy Pulsar

Among the 27 associations with known pulsars, we find 20 with significant pulsations above 10 GeV, and 12 with pulsations above 25 GeV, suggesting that the Crab pulsar will not remain the only pulsar to be detected by current and future IACTs.

The *Fermi*-LAT Collaboration, 2013

While over 200 pulsars are known to emit gamma rays in the MeV to GeV regime, only two pulsars have been detected at Very High Energy (VHE, $\gtrsim$100 GeV) so far: the Crab pulsar in the Northern Hemisphere and the Vela pulsar in the Southern Hemisphere. The reason behind this difference is that the vast majority of gamma-ray pulsars seem to exhibit a spectral cut-off at around a few GeV, as discussed in Sect. 2.4. However, as we saw in Chap. 5, in the case of the Crab pulsar the spectrum continues far beyond these energies up to $\sim$1 TeV. The Vela pulsar is the brightest steady source of gamma rays in the sky and it is still under debate whether its detection above 100 GeV is compatible with the cut-off seen by *Fermi*-LAT (see Fig. 7.1). The exact spectral shape of Vela's VHE component is still unknown and so far the Crab pulsar is the only pulsar that exhibits a power-law-like VHE tail extending far into the TeV regime.

From the experimental point of view the spectral breaks of gamma-ray pulsars fall in a rather unfortunate energy range, as indicated in Fig. 7.1. Gamma rays above $\sim$100 MeV can be efficiently detected from space, for example by the Large Area Telescope (LAT) on board of the *Fermi* satellite ([1], see Sect. 1.1). Above $\sim$10 GeV, however, the sensitivity of space-borne detectors starts to decrease rapidly due to their limited size. From the ground, Imaging Atmospheric Cherenkov Telescopes (IACTs, see Sect. 1.1) are able to detect gamma rays in the GeV to TeV range, peaking in sensitivity at around $\sim$500 GeV. Current generation IACTs are capable of detecting gamma rays of energy as low as $\sim$30 GeV, but have to fight with an overwhelming background at those energies [2, 3]. Therefore, the exact shape of the

© Springer Nature Switzerland AG 2019
D. Carreto Fidalgo, *Revealing the Most Energetic Light from Pulsars and Their Nebulae*, Springer Theses, https://doi.org/10.1007/978-3-030-24194-0_7

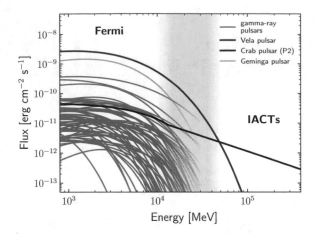

Fig. 7.1 Sketch of 60 gamma-ray pulsar spectra taken from the second *Fermi*-LAT Pulsar Catalog (2PC, [4]). The Crab pulsar (red line) and the Vela pulsar (blue line) are the only pulsars detected by IACTs so far. The second brightest pulsar in the *Fermi*-LAT sky is Geminga (green line) and has always been considered as a good candidate for follow-up observations by IACTs. Unfortunately, the spectral breaks of gamma-ray pulsars tend to fall in the energy range (a few tens of GeV, grey shaded area) where *Fermi*-LAT and IACTs perform suboptimally (see text). Data taken from [4] and this work (Color figure online)

spectral break and the maximum photon energy emitted by gamma-ray pulsars, both important ingredients for understanding the emission mechanism at work, are still poorly known.

In this chapter we report on a project conducted at the University of Hong Kong (HKU) during a research internship under the supervision of Dr. Pablo Saz Parkinson, a member of the *Fermi*-LAT collaboration, as well as Dr. Stephen C. Y. Ng and Prof. K. S. Cheng. Preliminary results were presented at the end of the internship at the 19th Annual Conference of the Physical Society of Hong Kong.[1] The goal of the project was straightforward: finding the next very-high-energy gamma-ray pulsar. With its overlapping energy range and wide field of view covering $\sim$20% of the sky at any time, *Fermi*-LAT is perfectly suited to provide observational guidance to IACTs that typically feature narrow field of views of about $\sim$4°. Therefore, looking through the wealth of gamma-ray pulsars discovered by *Fermi*-LAT is a good starting point to propose candidates for follow-up observations with IACTs.

A natural criterion to identify promising VHE candidates is the maximum energy of the *Fermi*-LAT photons that can be associated with the candidate pulsar. In the second catalog of hard *Fermi*-LAT sources (2FHL), Ackermann et al. [5] analyzed 80 months of *Fermi* data including only events with a reconstructed energy above 50 GeV. They were able to detect 360 sources with a test statistic (TS) greater than 25 ($\sim$5σ) and with at least 3 associated photons. But only one of the sources could be identified as a pulsar, that is the Vela pulsar. The approach of our project is to

---

[1] https://pshk.org.hk/, last accessed 20/06/2017.

specifically look for VHE photons around already known gamma-ray pulsars instead of scanning the entire sky. In contrast to the 2FHL we not only make use of the spatial information of the events but also incorporated the timing information to help with the association, which boosts our sensitivity. In fact, just two photons can already reveal significant pulsation if they fall right into the expected phase regions of the pulsar light curve (see, for example, the case of J1836+5925 in [6]).

In the first section of this chapter, we will give details about the data processing pipeline of our project focusing on how we incorporated the timing information of the events. In the second section we will explain why a selection of a rather small sample of pulsars was necessary and introduce each of the candidates of our study. In addition we will formulate the statistical treatment for claiming evidence of VHE pulsation. Finally, in Sect. 7.3 we will conclude the chapter with a presentation of the obtained results and a generic discussion.

7.1 Analysis Pipeline

The analysis pipeline can be divided into two major steps: the model analysis of the public *Fermi*-LAT data and the assignment of pulse phases to the events.

7.1.1 Model Analysis

For the first step we customize *Enrico*[2] [7], a community-developed Python package to simplify the *Fermi*-LAT analysis, which makes extensive use of the *Fermi Science Tools* (version V10R0P5) provided by the *Fermi* collaboration.[3] The high-level data analysis of *Fermi*-LAT data encounters two main issues [8]:

1. The large point spread function (PSF) of the instrument leads inevitably to source confusion, especially near and in the Galactic plane. Hence, every technique that is based on simply assigning photons to a source will most likely deliver poor results.
2. Detecting and characterizing gamma-ray sources usually requires integrations of one year or more. Since the orientation of the instrument with respect to the source is constantly changing, the instrument response function (IRF) is highly time dependent.

Out of these considerations the standard *Fermi* analysis chain employs a likelihood technique that naturally connects model and instrument parameters. The expected counts r of photons from a modeled source j and integrated over a time period t, can be computed as

---

[2]https://github.com/gammapy/enrico/.
[3]https://fermi.gsfc.nasa.gov/ssc/data/analysis/documentation/.

$$r_j\left(E', \mathbf{\Omega}'; \lambda\right) = \iiint dE d\Omega dt \, F_j\left(E, t, \mathbf{\Omega}; \lambda\right) \times \mathrm{IRF}\left(E, E', \mathbf{\Omega}, \mathbf{\Omega}'\right) \qquad (7.1)$$

where $F_i\left(E, t, \mathbf{\Omega}; \lambda\right)$ is the photon flux of the source model containing assumptions about the spatial and spectral shape (denoted as λ), and $\mathrm{IRF}\left(E, E', \mathbf{\Omega}, \mathbf{\Omega}'\right)$ is the instrument response function that connects the reconstructed energy E' and reconstructed event direction $\mathbf{\Omega}'$ with their true values E and $\mathbf{\Omega}$, respectively. Summing up over all modeled sources and assuming Poisson statistics, the maximization of the likelihood then tries to fit the expected counts from the model to the observed counts by varying the model parameters λ. For a detailed description of this procedure the reader is referred to Ackermann et al. [9].

For our analysis we use approximately 90 months of *Fermi*-LAT data processed with the PASS8 IRFs [10]. In a first iteration we select all events within a region of 15° centered at the position of our candidate pulsar between 100 MeV and 300 GeV. Then we apply the standard spatial and energy binning, as well as the standard quality cuts given by *Enrico*. For our initial fitting model we include all the sources listed in the 3rd *Fermi*-LAT Source Catalog (3FGL, [11]) with a distance of less than 17° to our candidate pulsar. We fix all model parameters to their catalog values for sources that are farther away than 5° (see Fig. 7.2). For sources within 5° we leave the spectral parameters free. The normalization of the isotropic background as well as the diffuse Galactic emission are also allowed to vary. After the maximization of the likelihood, we exclude all sources from our model that yield a test statistic of

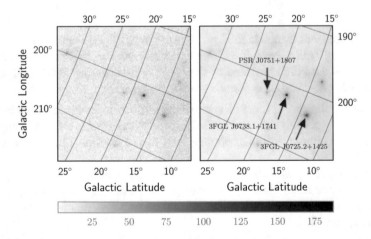

Fig. 7.2 Illustration of the *Fermi*-LAT model analysis. Both sky maps show a patch of the sky of roughly 15° × 15° around PSR J0751 and include photons from 100 MeV to 300 GeV. The grey scale denotes the number of events per spatial bin. **Left**: The observed photon counts map shows several bright point sources, isotropic background emission and some diffuse emission from the Galactic plane. **Right**: The model counts map incorporates all significant point sources from the 3FGL plus the diffuse emission. Apart from our source we annotate two nearby bright point sources corresponding to two blazars. The model that best fits the observed counts map is used in our analysis to calculate the photon weights (see text)

less than 1 ($\sim 1\sigma$) and repeat the maximization. The resulting model is then used to calculate and assign weights to each of the photons within 2° of our candidate. The 95% containment angle of *Fermi*-LAT at 10 GeV is around ~ 1° for a point source, hence the extraction radius of 2° should be sufficient to safely catch all potential VHE photons from the candidate. The assigned weight w_i represents the probability that the ith event originated from the candidate pulsar, computed by means of Eq. 7.1 as

$$w_i = \frac{r_{\text{psr}}\left(E'_i, \mathbf{\Omega}'_i\right)}{\sum_j r_j\left(E'_i, \mathbf{\Omega}'_i\right)}, \tag{7.2}$$

where E'_i and $\mathbf{\Omega}'_i$ is the reconstructed energy and position in the sky of the event, respectively. Thus, the first major step in our analysis chain leaves us with an event list of weighted photons for each candidate pulsar.

7.1.2 Pulsar Timing

In the second major step we want to include the timing information of the events by assigning pulse phases to each of the photons (see Appendix B for details). For this purpose we use the TEMPO2 package [12] together with the *Fermi* plug-in developed by Lucas Guillemot.[4] The required pulsar ephemerides are taken from the second *Fermi*-LAT Pulsar Catalog (2PC, [4]). However, the catalog is based on only 3 years of *Fermi*-LAT data until August 2011, while our data set includes data up to May 2016. Therefore, we have to make sure that the ephemerides are also valid for the newest data and if necessary update their parameters. For this procedure we follow the methods described in Kerr et al. [13] and Pletsch and Clark [14].

Normally each ephemeris comes with a *start* and *end date* that mark the time period for which the timing model correctly predicts the pulse phase up to a certain precision. To judge if the timing model is still effective beyond this period, we construct a pulse template out of the weighted pulsar light curve including only events for which the 2PC ephemeris is still valid (see Fig. 7.3, left plot). The template t_{pdf} consists of an arbitrary number m of skewed normalized Gaussians g_n with different amplitudes a_n plus a constant c:

$$t_{\text{pdf}}(\phi; \mathbf{u}) = c + \sum_{n=1}^{m} a_n \cdot g_n(\phi, \mathbf{u_n}), \tag{7.3}$$

where ϕ denotes the pulse phase. To be able to use t_{pdf} as a probability density function (PDF), we define the overall normalization such that $c + \sum_{n=1}^{m} a_n = 1$. The template parameters $\mathbf{u}$, like positions, amplitudes and widths of the Gaussians, are obtained by performing an unbinned likelihood fit evaluating

---

[4]https://fermi.gsfc.nasa.gov/ssc/data/analysis/user/Fermi_plug_doc.pdf, last accessed 26/06/2017.

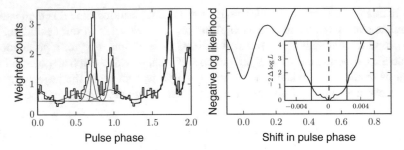

Fig. 7.3 Illustrating the method to determine the residuals of the timing models. **Left**: By way of example, we take the pulse profile of PSR J0340 using almost 4 years of data, plot two cycles and overlay our resulting fit. In the first cycle we show the single components (black thin lines) of the total fit function (see text and Eq. 7.3), in the second cycle we show the obtained pulse template (black thick line). **Right**: Using the template obtained in the left plot, we try to minimize the negative of the likelihood function defined in Eq. 7.4 by shifting the template over the pulsar light curve from a defined time period (see text). Once we find the global minimum of the likelihood we shift the template to the right and left (see inset) until the likelihood ratio test, $-2\Delta \log L$, gives us a test statistic of 1 (horizontal line), corresponding to 1σ error bars. The center (dashed line) of the obtained interval (solid vertical lines) is then taken as the residual of our timing model. We adopt the center of the interval, instead of the extremum of the likelihood, to ensure symmetrical error bars [13]

$$\log L(\mathbf{u}) = \sum_i \log \left[w_i \, t_{\mathrm{pdf}}(\phi_i; \mathbf{u}) + (1 - w_i) \right], \qquad (7.4)$$

where we incorporate the weights w_i of the photons.[5] The resulting template is then used to search for shifts in the light curve during our entire observation period. For this task we phase the whole data set with the given 2PC ephemerides and divide it into smaller time periods of 150 days. Depending on the signal strength of the pulsar, which we evaluate using the weighted H-test [15, 16], these periods are uniformly extended until the average significance of the signal yields at least $\sim 3\sigma$. In each of the time periods we slide our template over the corresponding light curve to look for the minimum of the negative log likelihood defined in Eq. 7.4 (see right panel of Fig. 7.3). The detected shifts with respect to the initial template can be interpreted as residuals and can be plotted versus time to probe the period of validity of the timing model. We note that in our case the initial timing solution was always good enough to roughly track the pulsar's rotation and conserve the basic pulse profile. Large deviations from the initial solution or sudden glitches (see Appendix B), however, can lead to significant broader peaks and completely smear out the signal in the phaseogram.

Once we observe a significant shift of the pulse profile with time we try to update the parameters of the ephemeris. Instead of constructing time of arrivals (TOAs) of pulses and try to fit the timing model to those TOAs (see, for example, [17]), we adopt

---

[5]For this fitting procedure we make use of the *GeoTOA* tool developed by M. Kerr and P. Ray (https://fermi.gsfc.nasa.gov/ssc/data/analysis/user/GeoTOA_README.txt, last accessed 02/04/2018).

an *event-based* timing as described in Pletsch and Clarke [14]. The pulse phase of each photon $\phi(t_i, \mathbf{v})$ is a function of the photon's arrival time t_i and the vector $\mathbf{v}$ that collects all the ephemeris parameters, like for example pulsar spin, position or orbit in a binary. If we fix the template t_{pdf} we can write Eq. 7.4 as

$$\log L(\mathbf{v}) = \sum_i \log \left[w_i \; t_{pdf}(\phi(t_i, \mathbf{v})) + (1 - w_i) \right] . \tag{7.5}$$

We can now fit the pulse phases of all events in our observation period to the given pulse profile t_{pdf} by varying $\mathbf{v}$. Instead of applying minimization algorithms to the negative of Eq. 7.5, we compute Bayesian posterior distributions for the relevant parameters by employing the sampler for Markov chain Monte Carlo *emcee*[6] [18]. Sampling in general is more efficient when encountering challenging likelihood surfaces in many dimensions, and the *emcee* implementation is particularly useful for sampling badly scaled distributions. We choose linearly uniform priors for all parameters and take the resulting 50th percentiles of the posterior distributions as our updated $\mathbf{v}$ of the ephemeris. To calculate the pulse phases during the sampling we refrain from using TEMPO2 but employ a new software package called PINT[7] (PINT Is Not TEMPO3). This package is still under development but is already able to produce residuals that agree with TEMPO2 to within $\sim$10 ns. It is mainly based on Python and modern libraries, which makes it very easy to implement in custom Python scripts.

In the project we only managed to partially automatize the whole analysis pipeline. The model analysis of the public *Fermi*-LAT data was done in an automated manner using the research computing facilities of the Hong Kong University. Each candidate pulsar, however, required a dedicated treatment regarding its timing analysis. Timing analysis in general is an iterative procedure. In a first sampling run we hold the pulsar's position and possible proper motion fixed to the 2PC values and only let the spin and orbital parameters vary. From this solution we refine the pulse profile template t_{psr} and in a subsequent sampling we also try to fit the sky position plus further parameters. These iterative steps and the necessity for an individual shape of each timing model (regarding the number of model parameters), make it difficult to fully automatize the timing analysis. To keep computing times at reasonable levels, we apply a minimum weight cut for the whole timing analysis that leaves us with roughly 5000 photons per candidate For dim pulsars with a strong background more photons are necessary, for bright and almost background free pulsars fewer photons are sufficient.

---

[6]http://dan.iel.fm/emcee/current, last accessed 26/06/2017.
[7]https://github.com/nanograv/PINT, last accessed 27/06/2017.

7.2   Selecting Candidates and Evaluating VHE Pulsation

For our project we focused on millisecond pulsars (MSPs) because by default they exhibit a more stable rotation, and therefore are easier to time requiring less parameters for their ephemeris. In addition the emission model by Harding and Kalapotharakos [19] distinguishes MSPs as promising candidates for synchrotron self-Compton emission at very-high energies, albeit at a challenging low flux level.

Out of the 40 MSPs in the second *Fermi*-LAT Pulsar Catalog we selected those pulsars observable from the MAGIC site at zenith angles below 30°, where MAGIC achieves the lowest energy threshold. This left us with 17 candidate pulsars. Since it is generally believed that millisecond pulsars have been spun up through accretion of matter from a companion star, most of them are found in binary systems. At the time of the project the software package PINT only supported the binary model by Blandford and Teukolsky [20]. Therefore, we also had to discard pulsars in binaries for which the 2PC ephemeris used a different model. These selection criteria and technical constraints left us with 3 binary and 3 isolated MSPs, which we introduce below. Furthermore we included Geminga and the millisecond pulsar PSR J0614-3329, a promising VHE candidate as pointed out by Pablo Saz Parkinson at the 6th International Fermi Symposium[8] in 2015.

PSR J0030+0451   PSR J0030 belongs to the less common class of isolated millisecond pulsars and was first discovered in an Arecibo drift scan search by Lommen et al. [21]. With a distance of just about 0.34 kpc [22],[9] it is one of the brightest gamma-ray MSP seen by *Fermi*-LAT. For MAGIC it is observable at a minimum zenith angle of 24°.

PSR J0102+4839   Targeted in a search for radio pulsation from unidentified *Fermi*-LAT sources, this binary MSP was discovered by Hessels et al. [23] in a Green Bank Telescope survey. The minimum zenith angle for MAGIC observations would be 19°.

PSR J0218+4232   This radio source was first identified as a binary MSP by Navarro et al. [24] at Jodrell Bank. Both radio and gamma-ray pulse profiles exhibit broad peaks. This, together with a large fraction of unpulsed radio emission, suggests that PSR J0218 is an aligned rotator. The minimum zenith angle for MAGIC observations would be 14°.

PSR J0340+4130   PSR J0340 was also discovered in a study of unidentified *Fermi*-LAT sources [23]. Like PSR J0030 it belongs to the class of isolated millisecond pulsars. For MAGIC it is observable at a minimum zenith angle of 13°.

PSR J0614-3329   PSR J0614 is the brightest MSP seen by *Fermi* in terms of integrated energy flux. Ransom et al. [25] discovered this binary MSP in a search for radio pulsars from unassociated sources in the *Fermi*-LAT Bright Source List [26] with the Green Bank Telescope. It is the only MSP that showed a strong hint of

---

[8]https://fermi.gsfc.nasa.gov/science/mtgs/symposia/2015/program/wednesday/session12B/PSaz Parkinson.pdf, last accessed 28/06/2017.

[9]http://www.atnf.csiro.au/research/pulsar/psrcat/, last accessed 28/06/2017.

pulsed emission above 25 GeV in the first *Fermi*-LAT catalog of sources above 10 GeV (1FHL, [6]). For MAGIC, however, this pulsar is only visible at very large zenith angles above 60°.

PSR J0633+1746 Known as *Geminga*, this middle-aged canonical pulsar is one of the brightest steady gamma-ray sources in the sky. Although various attempts have already been made by IACTs to detect it [27, 28], we applied our pipeline on this source as a first test. MAGIC can observe Geminga at a minimum zenith angle of 11°.

PSR J0751+1807 This binary pulsar was discovered in a search for radio pulsation from unidentified EGRET sources by Lundgren et al. [29]. It is one of the few MSPs that exhibit a pulse profile with 3 peaks. The minimum zenith angle for MAGIC observations would be 11°.

PSR J1939+2134 PSR J1939 was the first ever detected millisecond pulsar and to date remains the second fastest-spinning pulsar with a rotation period of only 1.56 ms. It belongs to the class of isolated MSPs and was discovered by Backer et al. [30]. For MAGIC it is observable at a minimum zenith angle of 7°.

To test for the presence of pulsations at very high energies, we follow the approach by Ackermann et al. [6]. Instead of relying on the weighted H-test, they used a likelihood ratio test (LRT), comparing the phase distribution of VHE events with the pulse profile at lower energies. This approach is naturally more sensitive, since contrary to the H-test, it involves assumptions about the expected pulse profile and thus benefits from additional information from the low-energy light curve, even if the profile may not necessarily be exactly the same as the high-energy profile. We extract the additional information from the weighted light curves above 1 GeV of our candidates, which we fit by an arbitrary number of skewed Gaussians to obtain the low-energy PDF $p_{LE}(\phi)$ as described in Eq. 7.3 (see Fig. 7.4). Following Eq. 2 in [6] we then write the PDF for the high-energy photons as:

$$p_{HE}(\phi) = (1 - x) + x \cdot \frac{p_{LE}(\phi) - c}{1 - c}, \tag{7.6}$$

where the parameter $x \in [0, 1]$ controls the resemblance of p_{HE} to p_{LE} ($x = 1$) or to a uniform phase distribution ($x = 0$, no pulsation). c represents the unpulsed component of the pulse profile as described in Eq. 7.3. The high-energy photons are extracted within a radius of 0.5° around the candidate for front-converting events and 1.0° for back-converting events, roughly corresponding to the 95% containment angle of the reconstructed incoming photon direction at 10 GeV [10]. Since we do not want to make any assumptions on the spectral shape of the candidate pulsars at the highest energies, we refrain from applying weights to the high-energy photons (see Eq. 7.2). For the likelihood ratio test we maximize the likelihood $L(x)$ derived from p_{HE} with respect to x and compare it to the null hypothesis $x = 0$, that there is no pulsation. By construction, the null hypothesis yields a likelihood of 1 and the test statistic TS $= -2 \ln \left(L(0)/L(\hat{x}) \right)$ can be simplified to TS $= 2 \ln L(\hat{x})$. Following [6] we convert the TS value to a p-value P and set a threshold of $P < 0.05$ to claim evidence for pulsation.

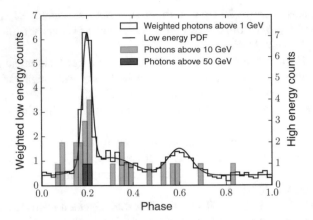

Fig. 7.4 Simulated light curves for an imaginary pulsar to illustrate our VHE pulsation test. From the weighted and normalized pulse profile above 1 GeV (black step histogram) we construct our pulse template (black line) and use it to test if the unweighted high-energy photons (for example above 10 GeV, red histogram, and 50 GeV, blue histogram) follow the same PDF or rather a uniform phase distribution (see Eq. 7.6) (Color figure online)

7.3 Results and Discussion

As discussed in the previous section, we have to make sure that we correctly predict the pulse phase for each photon to evaluate a possible pulsation at the highest energies. In Fig. 7.5 we show the results of our timing analysis by plotting the phase shifts versus time, which can be interpreted as the residuals of our timing models. For each candidate we achieved a root-mean-square error (RMSE) of smaller than 20 milliperiods (mp) and an improvement of at least 15% compared to the outdated timing models. The smallest RMSE are obtained for PSR J0030 and PSR J0633 yielding an average timing residual of 0.8 and 0.9 mp, respectively, for a time span of 150 days. This is not a surprise since both belong to the 4 brightest millisecond pulsars seen by Fermi in terms of photon fluxes, and more importantly, have sharp peaks which allow for a precise timing. In contrast, due to its broad peak, PSR J0218 could only be timed with an accuracy of 17.7 mp although it is brighter than PSR J0030. For the by far brightest source in our sample, Geminga, we obtained an average timing residual of 2.5 mp integrated over a time period of 150 days, which is sufficiently small for our studies. To achieve an even better timing solution, one would have to add further parameters to the ephemeris to cope with the higher timing noise that canonical pulsars generally exhibit compared to millisecond pulsars. In the case of PSR J1939 the weak signal and strong background required a longer integration time of 200 days to obtain a feasible pulsar light curve for the fits.

Having ensured the correct timing of our candidates over the entire observation period of 6 years we went on to search for pulsed emission at energies above 10, 25 and 50 GeV. Figure 7.6 shows the pulsar light curves for our candidates at the

Fig. 7.5 Residuals of our timing models versus time. The residuals, or phase shifts that are necessary to align the template to the light curve (see text), are given in milliperiods. Black squares show the residuals of the ephemerides taken from the second *Fermi*-LAT Pulsar Catalog (2PC). Most of them are valid up to ~56,000 MJD. For PSR J0614 and PSR J0633 (Geminga) we started from newer ephemerides given by the *Fermi*-LAT collaboration at https://confluence.slac. stanford.edu/display/ GLAMCOG/LAT+Gamma-ray+Pulsar+Timing+Models (last accessed 09/07/2017). Red dots show the residuals of our updated timing models (see text). For each data point we integrated 150 days of data. Due to its weakness and high background, PSR J1939 required an integration of 200 days (Color figure online)

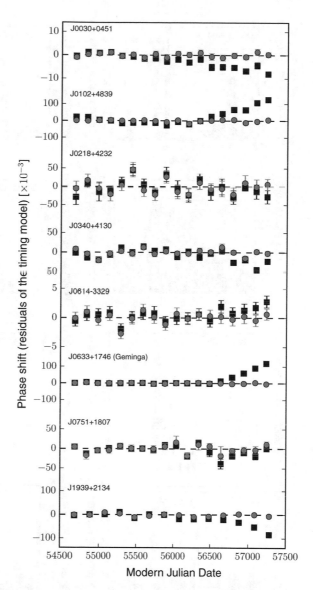

different energies and Table 7.1 summarizes the corresponding *p*-values for pulsed emission. All our candidates show evidence of pulsed emission above 10 GeV, except PSR J1939 whose *p*-value of 0.09 misses to pass the threshold of $P < 0.05$. Above 25 GeV only our two brightest sources, PSR J0614 and PSR J0633 (Geminga), are clearly detected at the ~5σ and ~6σ level, respectively, while PSR J0218 shows a hint of pulsation with $P = 0.073$. In the case of Geminga we see evidence of pulsation up to 37 GeV. The most energetic photon from the direction of Geminga

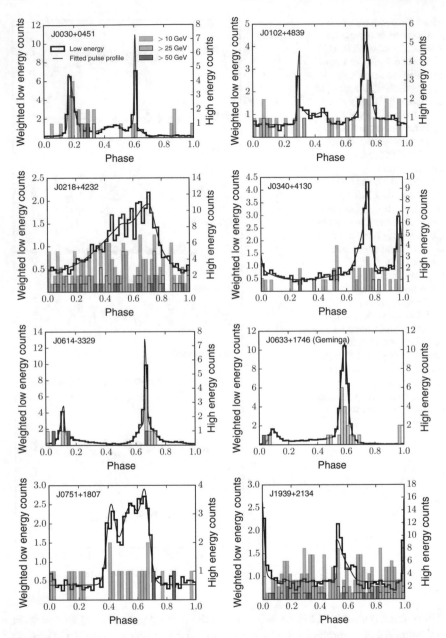

Fig. 7.6 Pulsar light curves of our candidates at different energies. The black step histogram (left y axes) shows the normalized weighted pulse profile for energies above 1 GeV, except for our brightest source, Geminga, for which we chose a low energy cut of 5 GeV. The solid black line depicts the fitted pulse profile of the low energy counts and is used as p_{LE} (see Sect. 7.2). The colored histograms (right y axes) show the unweighted light curves above following energies: red >10 GeV, green >25 GeV and blue >50 GeV. For our two brightest sources, PSR J0614 and Geminga, we omitted the light curve above 10 GeV (Color figure online)

Table 7.1   Results of the VHE pulsation tests

Pulsar	Period (ms)	$\dot{E}$	P_{10} (erg/s)	n_{25}	P_{25}	n_{50}	P_{50}
J0030	4.9	3.5×10^{33}	1.8×10^{-4}	3	>0.1	1	>0.1
J0102	3.0	1.8×10^{34}	1.6×10^{-2}	3	>0.1	0	>0.1
J0218	2.3	2.4×10^{35}	3.6×10^{-2}	48	7.3×10^{-2}	14	>0.1
J0340	3.3	7.8×10^{33}	2.5×10^{-5}	5	>0.1	1	>0.1
J0614	**3.1**	$\mathbf{2.2 \times 10^{34}}$	$\mathbf{<10^{-10}}$	**12**	$\mathbf{2.0 \times 10^{-6}}$	**3**	$\mathbf{2.6 \times 10^{-2}}$
J0633	**237.1**	$\mathbf{3.2 \times 10^{34}}$	$\mathbf{<10^{-10}}$	**23**	$\mathbf{3.4 \times 10^{-10}}$	**1**	**>0.1**
J0751	3.5	7.3×10^{33}	1.8×10^{-2}	2	>0.1	2	>0.1
J1939	1.6	1.1×10^{36}	>0.05	36	>0.1	12	>0.1

Notes The two bold rows correspond to the pulsars PSR J0614 and PSR J0633 (Geminga), which are firmly detected above 25 GeV. Columns 2 and 3 state the spin period and the spin-down power of the corresponding pulsar. Columns 4, 6 and 8 give the p-value of our VHE pulsation test (see text) above 10, 25 and 50 GeV, respectively. Columns 5 and 7 give the number of photons in the light curves above 25 and 50 GeV, respectively

has a reconstructed energy of $\sim$52 GeV, which, however, falls rather in the off-pulse phase region close to the first peak and is more likely due to unpulsed background emission. Above 50 GeV the light curve for PSR J0614 looks more compelling. The 3 photons with energies of approximately 61, 206 and 211 GeV yield a p-value of 0.026, and hence pass our threshold to claim evidence for pulsation. The low p-value is mainly due to the 61 GeV photon that falls right in the center of the second peak at phase $\sim$0.66. We validated the p-value with Monte Carlo simulations. Generating 10^5 random sets of 3 phases uniformly distributed between 0 and 1, we performed exactly the same test for VHE pulsation on each of these fake data sets and obtained a rate of false positives consistent with the p value. If we only consider the 61 GeV photon, the probability that a single photon from an unpulsed background falls right in the center of the second peak is actually less than 0.5% (2.8σ).

Summarizing our results, we can confirm the steep falling spectrum for Geminga after its break at $\sim$5 GeV which so far impeded the detection by IACTs at energies above 50 GeV (see for example, [28]). Whether the lack of VHE photons from Geminga is due to the limited sensitivity of current generation instruments or due to a spectral cut-off, is still under debate and will surely be investigated further in the future. From the observational point of view, a better candidate for VHE pulsation at the moment seems to be the millisecond pulsar PSR J0614-3329. Its light curve shows evidence of pulsed emission from the second peak up to 61 GeV and includes 2 intriguing $\sim$200 GeV photons around the first peak, albeit statistically insignificant when comparing their phase distribution to the pulse profile at 1 GeV. Our findings corroborate claims from a recent publication by Xing and Wang [31] who studied 39 MSPs with 7.5 years of *Fermi*-LAT data and managed to compute a spectrum up to 60 GeV for PSR J0614. With a declination of $-33°$, this candidate would be well suited for IACT observations by the H.E.S.S. telescopes in the Southern Hemisphere. For IACT observations the MSP PSR J0614 has two technical advantages when compared to young gamma-ray pulsars: (i) its location well outside of the Galactic plane

(Galactic latitude $b = -21.8°$), translates to a lower energy threshold for IACTs due to the reduced night sky background (NSB); (ii) pulsar wind nebulae are not expected for recycled pulsars, and hence eliminates a potential background component. The detection of PSR J0614-3329 by IACTs would ultimately reveal millisecond pulsars as VHE emitters and add a third member to the group of pulsars detected by IACTs.

References

1. Atwood WB et al (2009) The large area telescope on the Fermi Gamma-Ray space telescope mission. Astrophys J 697(2):1071–1102. https://doi.org/10.1088/0004-637X/697/2/1071
2. Holler M et al (2015) Observations of the Crab Nebula with H.E.S.S. Phase II. In: 34th international cosmic ray conference. The Hague, Netherlands, page Id 1046
3. Dazzi F et al (2015) Performance studies of the new stereoscopic sum-Trigger-II of MAGIC after one year of operation. In: 34th international cosmic ray conference. The Hague, Netherlands, page Id 608
4. Abdo AA et al (2013) The second Fermi large area telescope catalog of Gamma-Ray pulsars. Astrophys J Suppl Ser 208(2):17. https://doi.org/10.1088/0067-0049/208/2/17
5. Ackermann M et al (2016) 2FHL: the second catalog of hard Fermi-LAT sources. Astrophys J Suppl Ser 222(1):5. https://doi.org/10.3847/0067-0049/222/1/5
6. Ackermann M et al (2013) The first Fermi-LAT catalog of sources above 10 GeV. Astrophys J Suppl Ser 209(2):34. https://doi.org/10.1088/0067-0049/209/2/34
7. Sanchez DA, Deil C (2013) Enrico: a python package to simplify Fermi-LAT analysis. In: 33rd international cosmic ray conference, vol 3. Rio de Janeiro, Brazil, page 6361
8. Kerr M (2010) Likelihood methods for the detection and characterization of Gamma-ray pulsars with the Fermi large area telescope. Ph.D. thesis
9. Ackermann M et al (2012) The Fermi large area telescope on orbit: event classification, instrument response functions, and calibration. Astrophys J Suppl Ser 203(1):4. https://doi.org/10.1088/0067-0049/203/1/4
10. Atwood W et al (2013) Pass 8: toward the full realization of the Fermi-LAT scientific potential. arXiv
11. Acero F et al (2015) Fermi large area telescope third source catalog. Astrophys J Suppl Ser 218(2):23. https://doi.org/10.1088/0067-0049/218/2/23
12. Hobbs GB et al (2006) Tempo2, a new pulsar-timing package - I. An overview. Monthly Not R Astron Soc 369(2):655–672. https://doi.org/10.1111/j.1365-2966.2006.10302.x
13. Kerr M et al (2015) Timing Gamma-Ray pulsars with the Fermi large area telescope: timing noise and astronometry. Astrophys J 814(2):128. https://doi.org/10.1088/0004-637X/814/2/128
14. Pletsch HJ, Clark CJ (2015) Gamma-Ray timing of redback PSR J2339–0533: hints for gravitational quadrupole moment changes. Astrophys J 807(1):18. https://doi.org/10.1088/0004-637X/807/1/18
15. Kerr M (2011) Improving sensitivity to weak pulsations with photon probability weighting. Astrophys J 732(1):38. https://doi.org/10.1088/0004-637X/732/1/38
16. de Jager OC et al (1989) A powerful test for weak periodic signals with unknown light curve shape in sparse data. Astron Astrophys 221:180–190
17. Ray PS et al (2011) Precise Gamma-Ray timing and radio observations of 17 Fermi Gamma-ray pulsars. Astrophys J Suppl Ser 194(2):17. https://doi.org/10.1088/0067-0049/194/2/17
18. Foreman-Mackey D et al (2013) emcee : the MCMC Hammer. Publ Astron Soc Pac 125(925):306–312. https://doi.org/10.1086/670067
19. Harding AK, Kalapotharakos C (2015) Synchrotron self-compton emission from the crab and other pulsars. Astrophys J 811(1):63. https://doi.org/10.1088/0004-637X/811/1/63

20. Blandford R, Teukolsky SA (1976) Arrival-time analysis for a pulsar in a binary system. Astrophys J 205:580. https://doi.org/10.1086/154315
21. Lommen AN et al (2000) New pulsars from an arecibo drift scan search. Astrophys J 545(2):1007–1014. https://doi.org/10.1086/317841
22. Manchester RN et al (2005) The Australia telescope national facility pulsar catalogue. Astron J 129(4):1993–2006. https://doi.org/10.1086/428488
23. Hessels JWT et al (2011) A 350-MHz GBT survey of 50 faint Fermi γ-ray sources for radio millisecond pulsars'. In: Radio pulsars: an astrophysical key to unlock the secrets of the universe, pp 40–43. https://doi.org/10.1063/1.3615072
24. Navarro J et al (1995) A very luminous binary millisecond pulsar. Astrophys J, 455(1). https://doi.org/10.1086/309816
25. Ransom SM et al (2011) Three millisecond pulsars in Fermi LAT unassociated bright sources. Astrophys J 727(1):L16. https://doi.org/10.1088/2041-8205/727/1/L16
26. Abdo AA et al (2009a) Fermi/large area telescope bright gamma-ray source list. Astrophys J Suppl Ser 183(1):46–66. https://doi.org/10.1088/0067-0049/183/1/46
27. Aliu E et al (2015) A search for pulsations from geminga above 100 GeV with VERITAS. Astrophys J 800(1):61. https://doi.org/10.1088/0004-637X/800/1/61
28. Ahnen ML et al (2016) Search for VHE gamma-ray emission from geminga pulsar and nebula with the MAGIC telescopes. Astron Astrophys 591:A138. https://doi.org/10.1051/0004-6361/201327722
29. Lundgren SC et al (1995) A millisecond pulsar in a 6 hour orbit: PSR J0751+1807. Astrophys J 453:419. https://doi.org/10.1086/176402
30. Backer DC et al (1982) A millisecond pulsar. Nature 300(5893):615–618. https://doi.org/10.1038/300615a0
31. Xing Y, Wang Z (2016) Fermi study of γ-ray millisecond pulsars: the spectral shape and pulsed emission from J0614–3329 up to 60 GeV. Astrophys J 831(2):143. https://doi.org/10.3847/0004-637X/831/2/143

Part III
Looking for a Pulsar Wind Nebula in the Outer Part of Our Galaxy

Chapter 8
Does PSR J0631 Power a Pulsar Wind Nebula?

We find clear evidence that pulsars with large spin-down energy flux are associated with VHE γ-ray sources.

The H.E.S.S. Collaboration, 2007

Pulsar wind nebulae (PWNe) constitute the most numerous TeV gamma-ray sources in our galaxy. Most of the 19 firmly identified pulsar wind nebulae to date were spotted in the H.E.S.S. Galactic Plane Survey[1] (HGPS, [1]) from the Southern Hemisphere, but also MAGIC and VERITAS contributed to this population with the discoveries of 3C 58 and CT1 towards the outer part of the Milky Way [2, 3]. TeV pulsar wind nebulae are usually located close to a Galactic spiral arm structure, where the dense environment provides a natural birth place for pulsars (Fig. 8.1). In particular, the Scutum-Centaurus arm near the Galactic center hosts almost half of the current TeV PWNe population [4]. To increase the population with members situated in the outer part of our galaxy, MAGIC selected and observed 6 young and energetic gamma-ray pulsars that could potentially power TeV pulsar wind nebulae [5].

Here we will concentrate on one of these candidates, PSR J0631+1036, for which the author of this thesis was the main analyzer. The chapter will introduce the reader to the pulsar PSR J0631+1036, summarize previous observational results in the context of a TeV pulsar wind nebula and conclude by explaining why this pulsar was chosen as a candidate.

---

[1]https://www.mpi-hd.mpg.de/hfm/HESS/hgps/, last accessed 10/04/2018.

© Springer Nature Switzerland AG 2019
D. Carreto Fidalgo, *Revealing the Most Energetic Light from Pulsars and Their Nebulae*, Springer Theses, https://doi.org/10.1007/978-3-030-24194-0_8

Fig. 8.1 Locations of the so far firmly detected TeV pulsar wind nebulae in our galaxy (green dots, [4]) and PSR J0631 (red dot). Some of the nebulae closest to our solar system are labeled. Most of the firmly detected TeV PWNe are found on the Scutum-Centaurus arm near the Galactic center. Background image taken from http://www.eso.org/public/images/eso1339e/, last accessed 02/06/2017 (Color figure online)

8.1 The Young and Energetic Gamma-Ray Pulsar PSR J0631+1036

PSR J0631+1036 was discovered by Zepka et al. [6] in a radio search for counterparts of X-ray sources found in images of the Image Proportional Counter (IPC) on the *Einstein* Space Observatory [7]. It is interesting to note that its discovery was actually a chance coincidence of less than 1% since later on the X-ray source was found to be unrelated to the pulsar as we will describe further down. With a rotation period of $P = 288$ ms and a period derivative of $\dot{P} = 1.05 \times 10^{-13}$, PSR J0631 belongs to the young (<100 kyr) and energetic ($>10^{35}$ erg s^{-1}) group of canonical pulsars. We summarize some of the observed and derived radio timing parameters in Table 8.1. Zepka et al. [6] noted that PSR J0631 exhibits an unusual high dispersion measurement for its position near the Galactic anticenter and derived an estimated distance of 6.5 kpc following a Galactic electron density model by Taylor and Cordes [8].

Table 8.1 Observed and derived radio timing parameters for PSR J0631+1036

Parameter	Value
Right ascension (J2000) $\cdots\cdots\cdots\cdots\cdots$	$06^h 31^m 28^s$
Declination (J2000) $\cdots\cdots\cdots\cdots$	$+10°37'03''$
Galactic longitude, l $\cdots\cdots\cdots\cdots$	$201.2°$
Galactic latitude, b $\cdots\cdots\cdots\cdots$	$0.45°$
Period, P $\cdots\cdots\cdots\cdots$	0.287800021959 s
Period derivative, $\dot{P}$ $\cdots\cdots\cdots\cdots$	1.046741×10^{-13}
Dispersion measure, DM $\cdots\cdots\cdots\cdots$	125.36 cm^{-3} pc
Distance, d $\cdots\cdots\cdots\cdots$	2.1 kpc
Characteristic age, τ_c $\cdots\cdots\cdots\cdots$	4.36×10^4 yr
Spin-down power, $\dot{E}$ $\cdots\cdots\cdots\cdots$	1.7×10^{35} erg s^{-1}
Surface magnetic field strength, B $\cdots\cdots\cdots\cdots$	5.6×10^{12} G

Notes We adopt the values from the ATNF catalog [11]. Most measurements were conducted by Yuan et al. [12] and the distance has recently been updated following the electron density model by Yao et al. [10].

They argued, however, that much of the dispersion could be caused by the dark cloud LDN 1605 and ionized material associated with the star-forming region 3 Mon in the pulsar's foreground that was not taking into account in the electron density model, and therefore adopted a distance of 1 kpc throughout their paper. They motivated this value by its consistency with the amount of supposedly X-ray absorption and the efficiency of gamma-ray emission they allegedly found in EGRET data. Both claims turned out to be false and their assumed value of 1 kpc for the distance should be taken with care. More recent electron density models by Cordes and Lazio [9] and Yao et al. [10] yield distances of 3.6 ± 1.3 kpc and 2.1 ± 1.9 kpc, respectively, for which the errors of the latter are given as 95% confidence intervals. In the remainder of this work we will adopt a distance of 2.1 kpc, as also stated in the ATNF pulsar catalog[2] [11].

In radio PSR J0631 exhibits a rare four-component pulse profile with two pairs of nearly symmetrical peaks (see Fig. 8.2). The near even spacing of its four features, the deep emission minimum in the profile center and the weakness of its outer pair with respect to the inner pair distinguish its unusual profile from the rest of the four-component population. Teixeira et al. [13] show that such a profile can be explained by the widely used core-double-cone beaming model but needs an unrealistic fine tuning when assuming circular beams.

In the discovery paper Zepka et al. [6] associated the radio emission to the *Einstein* X-ray source 2E 0628.7+1037 despite a 75" discrepancy in their respective source positions. This discrepancy puts the radio position of PSR J0631 outside of the nominal 90% confidence area of the *Einstein* error circle, but Zepka et al. [6] argued that the shadowing of the support structure most likely led to an underestimation of

---

[2]http://www.atnf.csiro.au/research/pulsar/psrcat/.

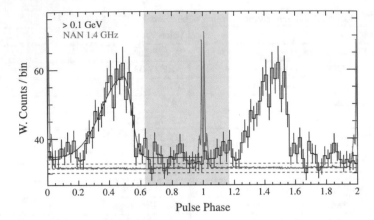

Fig. 8.2 Radio (Nançay Radio Observatory) and gamma-ray (*Fermi*-LAT) pulse profile of PSR J0631+1036. The broad gamma-ray peak lags the four-component radio pulse by ~0.5 in phase. The grey shaded patch depicts the off-phase region for the gamma-ray emission. Plot taken from Abdo et al. [14]

the error circle for the observed position in the sky. Therefore, they tried to crosscheck their results with an image from the ROSAT satellite [15], which confirmed the X-ray source with an estimated position 60" away from PSR J0631. Unfortunately also the ROSAT image was affected by a partial occultation from the detector supporting ribs, making it impossible to properly evaluate the error circle for the source position. The final piece of evidence to connect both sources to each other, seemed to be the detection of pulsed X-ray emission by Torii et al. [16] using data from the ASCA observatory [17]. Their modulation of the signal had the same period as the radio pulsar and they derived an observed pulsed fraction of 45% for the X-ray emission, similar to the one found in Geminga. In 2002, however, Kennea et al. [18] "set the record straight" by analyzing images from the XMM-*Newton* observatory, which provided a much improved spatial resolution [19]. With an angular resolution of around 6", it became clear that the radio position of PSR J0631, which was cross-checked using VLA data, is not consistent with the X-ray position of 2E 0628.7+1037, but in fact lies 75" further north. Kennea et al. [18] also searched for pulsed emission from the area around PSR J0631 but found no significant excess with the modulation of the pulsar's rotation period. Given the pulsed fraction claimed by Torii et al. [16], XMM-*Newton* should have detected the pulsed emission with high significance, and Kennea et al. [18] therefore conclude that the modulation detected by Torii et al. [16] is a statistical artifact. They instead associate the X-ray emission of 2E 0628.7+1037 with an A0 star 12 to 34 pc away from our solar system.

In 2010, Weltevrede et al. [20] announced the detection of pulsed gamma rays from PSR J0631 with *Fermi*-LAT. The pulse profile does not resemble the one claimed to be marginally found in EGRET data by Zepka et al. [6] and their flux estimation is an order of magnitude larger than that obtained by *Fermi*. The folded gamma-ray light curve of PSR J0631 features a single broad peak with a radio lag of around ~0.50 in phase (see Fig. 8.2). Its spectrum is well described by a power-law plus exponential cut-off with a cut-off energy at $E_c = (6 \pm 1)$ Ge, which is the typical spectral shape observed for most of the gamma-ray pulsars [14]. Assuming a distance of 1 kpc, Abdo et al. [14] derived a gamma-ray luminosity L_γ and efficiency $\eta = L_\gamma / \dot{E}$ in the 0.1–100 GeV energy band, which seem to fall rather in the lower percentiles of the corresponding $\dot{E}$ bin. If we correct these values for a distance of 2.1 kpc, PSR J0631 exhibits very typical values of $L_\gamma = 2.5 \times 10^{34}$ and $\eta = 0.14$ for gamma-ray pulsars with a spin-down power of $\dot{E} \simeq 10^{35}$ erg s^{-1}.

Recent optical observations by Mignani et al. [21] with the Gran Telescopio Canarias (GTC) revealed no optical emission at the 3σ level down to magnitudes of 27 and 26.3 in the Sloan g' and r' band, respectively.

8.2 Observational Hints of a TeV Pulsar Wind Nebula

After pulsar wind nebulae were established as an abundant class of strong VHE gamma-ray emitters [22], subsequent studies concluded that only young and energetic pulsars grow TeV pulsar wind nebulae that are bright enough to be detected by current Cherenkov telescopes [23]. Because of this connection, in 2007 VERITAS looked for a TeV pulsar wind nebula in the vicinity of PSR J0631 during their first year of operation with the full array [24]. In 13 hours of effective observation time they observed no significant excess and computed an integral flux upper limit (UL) above 300 GeV of 1.6×10^{-12} cm^{-2} s^{-1} assuming a spectral index of 2.5 and point-like emission.

The next observational results on PSR J0631 regarding the search for an associated TeV pulsar wind nebula came from Milagro. After the release of the *Fermi* Bright Source List (BSL) in 2009 [25], Abdo et al. [26] conducted a search of the Milagro sky map for spatial correlations with a subset of Galactic sources from the BSL. For this search they reanalyzed the full eight-year Milagro data set employing an optimized gamma-hadron separation. With the additional data and the improved analysis, they firmly detected multi-TeV emission for the first time from an highly extended region of $\sim 2°$ around the Boomerang and the Geminga pulsar. They further reported on five objects that showed a greater than 3σ excess, among them PSR J0631 with a significance of 3.7σ (see Fig. 8.3). According to Abdo et al. [26] the probability of a single 3σ false positive in their analysis is of the order of $\sim 4.4\%$. Given the significance, they estimated a flux level of 4.7×10^{-16} at 35 TeV from the position of PSR J0631, but did not state any limits regarding a possible extension.

Motivated by the Milagro hot spot, *Fermi* included PSR J0631 in a search for GeV counterparts to possible TeV pulsar wind nebulae [27]. While this study focused on the energy band above 10 GeV, where the nebulae tend to become brighter than their gamma-ray pulsars, *Fermi* also performed complementary searches in their entire energy range by analyzing only events from the off-phase region of the gamma-ray pulsars [14, 28]. In all their searches they could not find significant GeV gamma-ray emission from a possible PWN around PSR J0631, obtaining a flux upper limit of

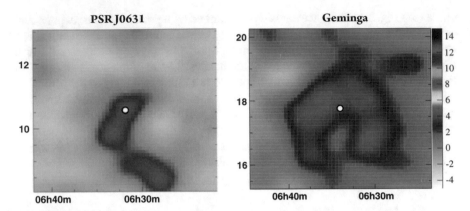

Fig. 8.3 Milagro sky maps for the region around PSR J0631 and the Geminga pulsar (as a reference for a firm detection). The white dots indicate the respective gamma-ray pulsar position as seen by *Fermi*-LAT. Both images were smoothed by a Gaussian of varying width ($0.4° - 1.0°$) depending on the expected angular resolution of the instrument at the position in the sky. PSR J0631 shows a hint of point-like emission at the 3.7σ level. For Geminga the significance is only 3.5σ, but increases to 6.3σ assuming extended emission with a $1°$ Gaussian profile. Horizontal axes denote right ascension, vertical ones show declination. The color scale indicates the statistical significance expressed in standard deviations. Images taken from Abdo et al. [26] (Color figure online)

0.4×10^{-10} cm^{-2} s^{-1} in the energy range between 100 and 316 GeV assuming a point source.

The results obtained with Milagro were recently questioned by the release of the second HAWC Observatory Gamma-ray Catalog [29]. With the improved sensitivity of the instrument and 17 months of data, HAWC should have detected the excess seen by Milagro at the position of PSR J0631 assuming a reasonable spectral index for the extrapolation of the flux level to lower energies (see Fig. 8.4).

The high spin-down power, young age and apparent proximity make PSR J0631 a prime candidate for powering a TeV pulsar wind nebula that could be detectable by current Cherenkov telescopes like MAGIC. The limits obtained by VERITAS above 300 GeV and the ones by *Fermi* in the energy range from 100 to ~300 GeV, are not stringent, especially considering that both ULs imply a point-like emission while the source is most likely extended given its recently updated distance of 2.1 kpc [10, 11]. Observing PSR J0631 for at least 30 hours, MAGIC should be able to significantly improve the limits and put to test the Milagro measurement even under the assumption of an extended emission region. It is fair to say that in the light of the recent HAWC results [29], MAGIC probably would have refrained from observing PSR J0631, since the Milagro hot spot was the main pillar of the observation proposals that were executed in the winter months of 2014/15 and 2015/16.

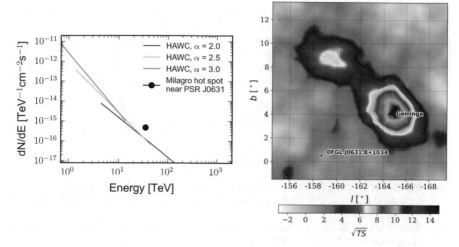

Fig. 8.4 Sensitivity curve and sky map from the HAWC observatory around PSR J0631. **Left:** The plot shows 5σ sensitivity curves of HAWC for a source declination of 10° and the flux measurement by Milagro for PSR J0631. HAWC states its sensitivity in function of the assumed spectral index α of a point source with a power-law like spectrum. The energy range of each sensitivity line defines the central 75% of the contribution to the test statistic (TS). **Right:** The TS around PSR J0631 and the Geminga pulsar as seen by HAWC with their 17 month data set. While the extended emission from Geminga seen by Milagro is confirmed, no significant excess is observed around PSR J0631. Horizontal and vertical axis show the Galactic longitude and latitude, respectively. The color scale indicates the significance given as the square root of the test statistic corresponding to standard deviations. Plots taken and reproduced from Abeysekara et al. [29] (Color figure online)

References

1. Abdalla H et al (2018a) The H.E.S.S. Galactic plane survey. Astron Astrophy 612:A1. https://doi.org/10.1051/0004-6361/201732098
2. Aleksić J et al (2014b) Discovery of TeV γ-ray emission from the pulsar wind nebula 3C 58 by MAGIC. Astron Astrophy 567:L8. https://doi.org/10.1051/0004-6361/201424261
3. Aliu E et al (2013) Discovery of TeV gamma-ray emission from CTA 1 by VERITAS. Astrophys J 764(1):38. https://doi.org/10.1088/0004-637X/764/1/38
4. Abdalla H et al (2018b) The population of TeV pulsar wind nebulae in the H.E.S.S. Galactic Plane Survey. Astron Astrophys 612:A2. https://doi.org/10.1051/0004-6361/201629377
5. MAGIC Collaboration et al (2018) MAGIC observations on Pulsar Wind Nebulae around high spin-down power Fermi-LAT pulsars. In preparation
6. Zepka A et al (1996) Discovery of three radio pulsars from an X-ray-selected sample. Astrophys J 456:305. https://doi.org/10.1086/176651
7. Giacconi R et al (1979) The Einstein /HEAO 2/ X-ray observatory. Astrophys J 230:540. https://doi.org/10.1086/157110
8. Taylor JH, Cordes JM (1993) Pulsar distances and the galactic distribution of free electrons. Astrophys J 411:674. https://doi.org/10.1086/172870
9. Cordes JM, Lazio TJW (2002) NE2001.I. A new model for the Galactic distribution of free electrons and its fluctuations
10. Yao JM et al (2017) A new electron-density model for estimation of pulsar and FRB distances. Astrophys J 835(1):29. https://doi.org/10.3847/1538-4357/835/1/29

11. Manchester RN et al (2005) The Australia telescope national facility pulsar catalogue. Astrophys J 129(4):1993–2006. https://doi.org/10.1086/428488
12. Yuan JP et al (2010) 29 glitches detected at Urumqi observatory. Mon Not R Astron Soc 354(1):811. https://doi.org/10.1111/j.1365-2966.2010.16272.x
13. Teixeira MM et al (2016) Investigations of the emission geometry of the four-component radio pulsar J0631+1036. Mon Not R Astron Soc 455(3):3201–3206. https://doi.org/10.1093/mnras/stv2520
14. Abdo AA et al (2013) The second Fermi large area telescope catalog of gamma-ray pulsars. Astrophys J Suppl Ser 208(2):17. https://doi.org/10.1088/0067-0049/208/2/17
15. Trümper J (1982) The ROSAT mission. Adv Space Res 2(4):241–249. https://doi.org/10.1016/0273-1177(82)90070-9
16. Torii K et al (2001) ASCA detection of pulsed X-ray emission from PSR J0631+1036. Astrophys J 551(2):L151–L154. https://doi.org/10.1086/320016
17. Tanaka Y et al (1994) The X-ray astronomy satellite ASCA. PASJ 46:L37–L41
18. Kennea J et al (2002) XMM-Newton sets the record straight: no X-ray emission detected from PSR J0631+1036
19. Aschenbach B (2002) In-orbit performance of the XMM-Newton x-ray telescopes: images and spectra, p 8. https://doi.org/10.1117/12.454367
20. Weltevrede P et al (2010) Gamma-ray and radio properties of six pulsars detected by the fermi large area telescope. Astrophys J 708(2):1426–1441. https://doi.org/10.1088/0004-637X/708/2/1426
21. Mignani RP et al (2016) Observations of three young γ-ray pulsars with the Gran Telescopio Canarias. Mon Not R Astron Soc 461(4):4317–4328. https://doi.org/10.1093/mnras/stw1629
22. Aharonian F (2005) A new population of very high energy gamma-ray sources in the Milky Way. Science 307(5717):1938–1942. https://doi.org/10.1126/science.1108643
23. Carrigan S et al (2007) Establishing a connection between high-power pulsars and very-high-energy gamma-ray sources. In: 30th International cosmic ray conference, Meirda, Mexico, pp 659–662
24. Aliu E et al (2008c) Search for VHE γ-ray emission in the vicinity of selected pulsars of the Northern Sky with VERITAS. In: AIP conference proceedings, pp 324–327. AIP. https://doi.org/10.1063/1.3076672
25. Abdo AA et al (2009a) Fermi/Large area telescope bright gamma-ray source list. Astrophys J Suppl Ser 183(1):46–66. https://doi.org/10.1088/0067-0049/183/1/46
26. Abdo AA et al (2009b) Milagro observations of multi-TeV emission from galactic sources in the fermi bright list. Astrophys J 700(2):L127–L131. https://doi.org/10.1088/0004-637X/700/2/L127
27. Acero F et al (2013) Constraints on the Galactic population of TeV pulsar wind nebulae using fermi large area telescope observations. Astrophys J 773(1):77. https://doi.org/10.1088/0004-637X/773/1/77
28. Ackermann M et al (2011) Fermi-LAT search for pulsar wind nebulae around gamma-ray pulsars. Astrophys J 726(1):35. https://doi.org/10.1088/0004-637X/726/1/35
29. Abeysekara AU et al (2017) The 2HWC HAWC observatory gamma-ray catalog. Astrophys J 843(1):40. https://doi.org/10.3847/1538-4357/aa7556

Chapter 9
MAGIC Observations of PSR J0631: Analysis and Results

For purposes of comparison with observations at high photon energies, I have computed the Compton-synchrotron spectrum of the Crab on the assumption that there exists therein production of a synchrotron spectrum extending from the radio to the optical to the x-ray region.

Robert J. Gould, 1965

As discussed in the previous chapter, the main argument of the observation proposal for PSR J0631, apart from being a prime candidate to exhibit a detectable TeV pulsar wind nebula, was the hot spot seen by Milagro at ~35 TeV. For this reason we were mainly interested in the higher energies accessible to MAGIC, for which observations under moderate moonlight are perfectly suitable [1]. Therefore, in the proposal we specifically asked for moon time observations with a zenith angle range up to 50°. In addition to being interested in the higher energies, most of the observation proposals for MAGIC require dark conditions to take advantage of the exquisite performance of the instrument at its lowest possible energy threshold, and hence observation time under moonlight is in general easier to obtain.

The chapter will describe the data set for PSR J0631 and focus on its analysis, which is slightly non-standard given the brighter light conditions during the observations and the possible extension of the source. We will conclude the chapter with a discussion of the results and review them in the context of the TeV pulsar wind nebulae population in our galaxy.

9.1 Data Set and Quality Selection

Due to its location near the Galactic anticenter, PSR J0631 is only observable for MAGIC during the winter months from October up to March. The observation campaign of PSR J0631 started in December 2014 and continued in December 2015 until March 2016, resulting in a total of ~43 h of data taken under moonlight of different phases and varying intensities in a zenith angle range from 15° to 50° (see Fig. 9.1).

© Springer Nature Switzerland AG 2019
D. Carreto Fidalgo, *Revealing the Most Energetic Light from Pulsars and Their Nebulae*, Springer Theses, https://doi.org/10.1007/978-3-030-24194-0_9

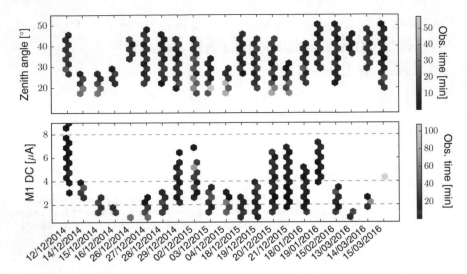

Fig. 9.1 Zenith angle distribution and medium anode current (DC) of the MAGIC-I camera for each observation night. The color scale denotes the respective observation time in minutes. **Top**: Our data was taken in a zenith angle range between approximately 15° and 50°, most of it below 30°. **Bottom**: We divide our data into 3 subsamples depending on the DC values of the MAGIC-I camera, which are a direct measure of the amount of background light (see text and Table 9.1). The selection criteria are depicted by the dashed horizontal lines (Color figure online)

To exclude data affected by adverse weather conditions or technical problems we used measurements of the LIDAR system and the Cloudiness parameter (see Sect. 3.3), and looked into the relevant run and logbooks of each observation night provided by the MAGIC collaboration. While additional background light does not affect the Cloudiness parameter, very strong moonlight can in principle disturb the LIDAR system. We made sure that our LIDAR measurements did not exhibit any unusual features that could have been caused by the increased background light level, and subsequently selected only data for which the LIDAR reported transmission values of more than 0.85 measured from 12 km above the telescopes. At the same time we removed all data with a Cloudiness above 0.40 as shown in Fig. 9.2. For time periods in which no LIDAR measurements were available we converted the Cloudiness parameter into a transmission value[1] and again applied a cut at 0.85. These selection criteria left us with around ∼37 h of good quality data that we used for the later analysis.

In case of data taken under moonlight, the event rate is only a suboptimal indicator of the data quality (see Sect. 3.4). The higher background light level caused by the moon leads to an increase of the individual pixel thresholds by the IPRC (see Sect. 3.3), and hence to an increased energy threshold and lower trigger rates. Therefore, low event rates do not necessarily imply adverse weather conditions but can also be a consequence of moonlight.

---

[1] As described in the bachelor thesis of Joel Betorz in 2015 at the Universitat Autònoma de Barcelona (UAB) under the supervision of Dr. Markus Gaug.

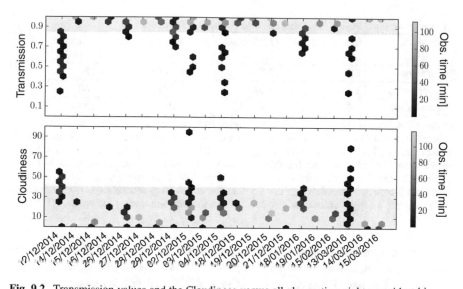

Fig. 9.2 Transmission values and the Cloudiness versus all observation nights considered in our analysis. The color scale denotes the respective observation time in minutes. **Top**: The transmission is measured from 12 km above the telescopes to the ground. For time periods where no LIDAR data is available, we derive an approximated transmission value from the Cloudiness parameter. Data with a transmission value above 0.85 is marked by a red shaded patch and will be used in the analysis. **Bottom**: At the same time we require a Cloudiness parameter below 0.4 (red shaded region). These selection criteria leave us with 36.9 h of good quality data out of the available ~43 h (Color figure online)

The data was taken after the change of the sampling speed of the MAGIC readout in November 2014 as mentioned in Sect. 3.3. During the observation campaign the instrument response did not change beyond the systematic uncertainties and the whole data set belongs to the same analysis period, namely ST.03.06.[2]

9.2 Moonlight Data and Extended Source Analysis

Moonlight during observations lead to an increase of background light (often referred to as Night Sky Background, NSB), and therefore to a higher noise level in the shower images of the telescopes. The instrument response changes and this has to be reflected in the MCs used for the analysis.

All our data was taken with nominal high voltages (HVs) applied to the PMTs, since the anode currents (DCs) were always well within the safety limits during the observations (Ahnen et al. [1] discuss alternatives when the moon is too bright for nominal HVs). To adapt the MCs for the higher background light level we divide

---

[2]For the naming convention see Sect. 5.1.

Table 9.1 Image cleaning parameters and size cuts for the subsamples

Subsample	Median DC [μA]	Mean/RMS of ped. [ph.e.]	Img. clean. levels [ph.e.]	Size cut [ph.e.]
Dim moon	<2	2.5/1.2	6.0/3.5	60
Moderate moon	2–4	2.9/1.3	7.0/4.0	80
Decent moon	4–8	3.6/1.5	9.0/5.5	150

Notes We define the subsamples given in Column 1 by the median DC values of the MAGIC-I camera shown in Column 2. The dim moon, moderate moon and decent moon sample contain approximately 10 , 16 and 11 h, respectively, of the total 37 h of observation time. Column 3 gives the mean and the RMS values for the photoelectron distribution of pedestal events in the MAGIC-I camera. Column 4 gives the values for the level 1 and level 2 parameter of the image cleaning, which are determined from the mean and RMS in Column 3. To guarantee a good match between MCs and data, we apply a minimum size cut stated in Column 5 following Ahnen et al. [1]

our data into three subsamples with respect to the median DC of all PMTs in the MAGIC- I camera, as shown in the lower plot of Fig. 9.1. The choice of MAGIC- I or II is arbitrary, but one has to keep in mind that a scaling is necessary when comparing the DCs of both cameras with each other [2]. The bin edges of the three subsamples are chosen following Ahnen et al. [1] and are based on internal studies by the MAGIC collaboration. We define our *dim moon* sample as data with a median DC of less than 2 μA, the *moderate moon* sample has DCs between 2 and 4 μA, and the *decent moon* sample contains data with up to 8 μA (see Table 9.1). The few data, approximately 0.5 h, with a median DC above 8 μA are discarded from the analysis. For each of the subsamples the data processing is done independently and the final analysis results are obtained by joining the high level products.

We process our data by means of the standard MAGIC analysis tool MARS (see Sect. 3.4 and [3]). In a first step the signals from the PMTs are converted into units of photoelectrons by means of interleaved calibration and pedestal runs. The resulting images are than cleaned using the sum-cleaning algorithm as described in Sect. 3.4. The increased cleaning levels for data with DCs above 2 μA make sure that fewer than ~10% of our shower images contain spurious patches of pixels (islands) caused by fluctuations in the NSB, which affect the event reconstruction (see Fig. 9.3). As a rule of thumb, to estimate the cleaning levels in advance one can set the first cleaning level to a 3σ fluctuation in the NSB, that is the mean value of photoelectrons in pedestal events plus 3 times their RMS, and the second level to a 1σ fluctuation. Higher image cleaning levels also imply a higher energy threshold, and hence their choice is always a compromise between keeping the number of additional islands in the shower images small, and achieving a low energy threshold. We estimate the energy thresholds of our subsamples by means of Eq. 1 in [1] and obtain thresholds of around 150, 210 and 280 GeV for our dim, moderate and decent moon sample, respectively, assuming a spectral index of 2.6. Since gamma-ray events with low sizes are harder to reconstruct and in general harder to distinguish from hadron-induced showers, the distribution of MCs and data for low sizes tend to show some discrepancies, which extend to higher sizes with an increasing level of background light. To ensure a good match

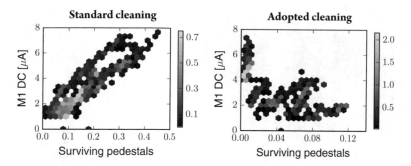

Fig. 9.3 The median DCs in the MAGIC-I camera versus the fraction of surviving pedestal events after the image cleaning. The color scale denotes the respective observation time in hours. **Right**: By applying standard image cleaning parameters, DC values above $\sim$2 μA lead to large fractions of surviving pedestal events, and hence to shower images with spurious islands. **Left**: Adopting the image cleaning parameters, the fraction of surviving pedestals drops significantly. For data above 4 μA the image cleaning seems a bit too conservative. Since we are mainly interested in the higher energies above 300 GeV, a low energy threshold is not needed and a generous cleaning will not have a negative impact on our analysis (Color figure online)

between data and MCs, a lower size cut is applied to every subsample depending on their background light level (see Table 9.1). We further introduce additional noise to the MCs for our moderate moon and decent moon subsample to mimic the effect of the moonlight. As described in detail in [1], the extra noise is injected right after the calibration of the simulated PMT signals, and is based on the charge distributions of pedestal events in our data, which obviously differ for both subsamples . We also add the same noise to our Off sample when training the random forest (RF, see Sect. 3.4 for details) for the gamma-hadron separation in the case of the moderate and decent moon sample, since it only contains observations performed under dark conditions. In contrast, for the dim moon sample the only difference to a standard MAGIC analysis is the slightly increased minimum size cut from 50 to 60 ph.e.

Following the standard MAGIC analysis chain (see Sect. 3.4), in a next step the cleaned shower images of both telescopes are parametrized by the classical *Hillas* ellipses and combined to obtain the stereoscopic parameters (see Sect. 3.2). Each event is tagged with a *Hadronness* by means of the RF technique, which is also used to reconstruct the event's arrival direction. The angular distance between the source position and the reconstructed event direction is expressed as Θ. The energy of the event is estimated with lookup tables built from the MC simulations corresponding to the respective subsample. We apply the same steps to a MC test sample to estimate the effective area of our analysis.

Normally MAGIC Monte Carlos are simulated for a point source 0.4° away from the camera center equal to the standard wobble offset (see Sect. 3.4). The MAGIC collaboration also provides *diffuse* MCs that are simulated in an uniform disc around the camera center with a radius of 1.5°. These MCs are typically used when the wobble offset of the observation is not the standard one or if one wants to analyze possible sources, as will be our case. When dealing with extended sources one has to take

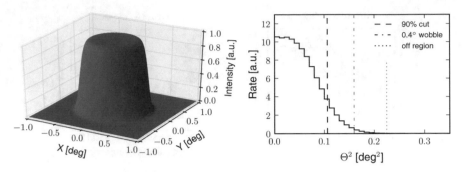

Fig. 9.4 Sketch of a brightness profile and the corresponding expected Θ^2 distribution under the assumption of a uniform acceptance of the camera and no background. **Left:** Shown is a disk-like brightness profile with $r = 0.3°$ in the camera plane centered at the camera center, convoluted with a Gaussian of $\sigma = 0.058°$ representing the PSF of the MAGIC telescopes. **Right:** The corresponding Θ^2 distribution. The 90% efficiency cut at $\sim 0.11°$ is denoted as a dashed line. The dash-dotted line shows the standard 0.4° wobble offset and the dotted line depicts the boundary to the antisource off region corresponding to the 90% efficiency Θ^2 cut

special care with contaminations of the background counts from the source itself. To estimate the maximum extension of a source that can be safely analyzed with the standard wobble offset of 0.4°, we perform some back-of-the-envelope calculations and plot the expected Θ^2 distribution. We convolute an assumed brightness profile with a 2 dimensional Gaussian with a sigma of $\sigma = 0.058°$ representing the PSF of our analysis above ~ 300 GeV.[3] The result of this convolution is plotted in the left panel of Fig. 9.4 for a disk-like brightness profile with radius $r = 0.3°$. Under the assumption of no background and an uniform acceptance across the whole MAGIC cameras [5], the expected θ^2 distribution is shown on the right in Fig. 9.4. For the Θ^2 cut, we aim for an efficiency of at least 90% to make sure that slight discrepancies between MCs and data regarding the source profile lead to an error smaller than MAGIC's systematic uncertainties of $\sim 15\%$ for the flux level. This results in a cut value of $\Theta^2 \simeq (0.33°)^2$ and consequently to a spill-over of source events into the off region of less than 0.01%, if we only consider the anti-source position to estimate the background (see Sect. 3.4 for details on the background estimation techniques in MAGIC). Hence, we conclude that a source exhibiting a disk-like brightness profile with a radius of up to 0.3° can be safely analyzed with the standard wobble offset of 0.4°.

For the last couple of years the MAGIC collaboration has put special effort into developing new tools for extended source analysis [6]. One of the new tools we will use in this analysis, is the *Donut MonteCarlo* executable that properly prepares the test MC sample for a correct estimation of the effective area in the case of extended sources. It convolves an arbitrary brightness profile with the energy dependent PSF,

---

[3]We estimate the PSF by analyzing Crab nebula data taken under similar conditions to our data. For this estimation we ignore the fact, that the tails of MAGIC's PSF can be slightly better reproduced by the King function or a double Gaussian, see for example [4].

takes into account the acceptance of the camera and recalculates the source position of each MC event to point to the center of the extended emission region.

As a last step in the analysis chain we apply signal extraction cuts to our final data and MC event lists. For the Θ^2 plot and the sky map shown in the next section, we use the *full range* cuts provided by the MAGIC collaboration, which are optimized for a wide energy range and contain a *Size* cut of 300 ph.e. (see Sect. 3.4). In case of the spectrum we apply energy dependent cuts by means of cut efficiencies calculated from the MC test sample. We require efficiencies of 90% for the *Hadronness* and the Θ^2 cut, and apply the size cuts specified in Table 9.1.

9.3   Analysis Results and Discussion

In the recently published population study of TeV pulsar wind nebulae, [7] define a *baseline model* and *varied models* to reflect the average trend of PWN evolution as well as the scatter of individual nebulae around the expectation. For a pulsar like PSR J0631 and a distance of 2.1 kpc, the baseline model predicts a PWN radius of $r_{pwn} \simeq 0.44°$ with a typical scatter from the varied models of a factor of $\sim$2 (see Table A.2 in [7]). With our observations, which were taken with the standard wobble offset of 0.4°, such a big extension cannot be tested without a significant leakage of gamma-rays from the source into the background region. However, since from the observational point of view the PWN extension of PSR J0631 is completely unknown, below we will give analysis results assuming a point-like emission and an extension of $r_{pwn} = 0.3°$, which is the maximum we can safely probe with our observational settings as discussed in the previous section.

Figure 9.5 shows the Θ^2 distributions of our On and Background events for the two assumptions, as well as sky maps of the test statistic (TS, see Sect. 3.4 for the exact definition). In case of point-like emission we estimate the background in the Θ^2 plot from 3 Off source positions, while in the case of extended emission, we only consider Θ^2 with respect to the antisource position. After defining a fixed Θ^2 cut, the significances of the excess are 1.4σ and 0.1σ, respectively, following ([8], Eq. 17) and taking into account trials. For the background estimation in the sky maps, we divide the camera plane along the wobble axes and only consider events coming from the half corresponding to the antisource position (see Sect. 3.4 for details). The maps are smeared using a Gaussian kernel with a sigma of 0.058° or 0.15°, depending on the assumed emission profile (point-like or a disk with a radius of 0.3°). The TS sky maps show no significant excess from the direction of PSR J0631 and the distribution of TS values in the field of view (FoV) is compatible with the null hypothesis of just background. Given the Milagro hot spot at 35 TeV, we repeat these detection exercises using the *high energy* cuts provided by the MAGIC collaboration, which are optimized for energies above 1 TeV. Again, we see no hint of a signal.

Following the non-detection, we compute upper limits (ULs) at the 95% confidence level applying the method by Rolke et al. [10]. We plot the differential ULs in Fig. 9.6, assuming a point-like emission and a disk-like emission profile with radius

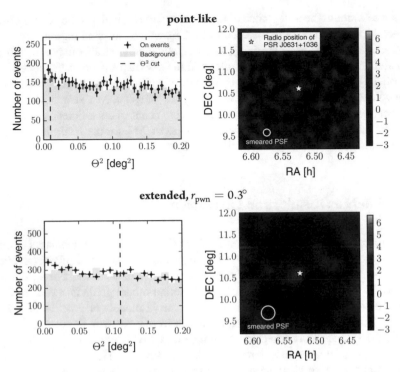

Fig. 9.5 Θ^2 distributions of On and Background events (left column), and the corresponding TS sky maps (right column). **Top row**: In the case of point-like emission the background in the Θ^2 plot is taken from 3 Off source positions and the Θ^2 cut (dashed vertical line) is fixed at 0.01 deg^2 $\simeq$ $(2\,\sigma_{psf})^2$. The resulting significance is 1.4σ for the observed excess. The TS sky map is smeared with a Gaussian kernel of $\sigma = \sigma_{psf}$. **Bottom row**: The background in the Θ^2 plot is estimated only from the antisource position to avoid possible contaminations from the source. We fix the Θ^2 cut to $r_{pwn}^2 + (2\,\sigma_{psf})^2 \simeq 0.11$ deg^2 obtaining a significance of 0.1σ. The corresponding TS sky map is smeared with a Gaussian kernel of $\sigma = 0.15°$ to enhance the significance of the possible extended emission (Color figure online)

$r_{pwn} = 0.3°$. In both cases we imply a power-law like spectrum with index -2.2, a typical value for TeV PWNe found in the study by Abdalla et al. [7]. Above 300 GeV we obtain integral flux upper limits of 6.0×10^{-13} cm^{-2} s^{-1} and 2.8×10^{-12} cm^{-2} s^{-1} for the two emission scenarios, respectively. For the discussion below we will adopt the more conservative limit, but note that an extension of the source larger than $r_{pwn} \gtrsim 0.5°$ could significantly affect our background estimation and render our upper limits too optimistic or even impede a detection. As a reference in Fig. 9.6 we also plot the spectrum of PSR J0631 as measured by *Fermi*-LAT below 100 GeV and include the flux estimation by Milagro at 35 TeV for their 3.7σ hot spot (see Sect. 8.2). If we extrapolate Milagro's estimation to lower energies under the assumption of a power-law with photon index 2.2, our integral upper limits are in tension with their results.

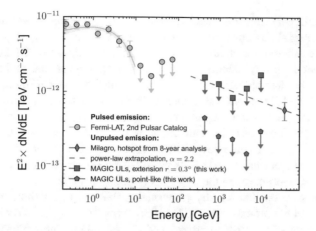

Fig. 9.6 Spectral measurements in the direction of PSR J0631+1036. The pulsed emission seen by *Fermi* LAT below 10 GeV is shown as yellow dots. The green diamond denotes the flux estimation of the Milagro hot spot (see text) and can be interpreted as steady emission from a possible pulsar wind nebula powered by PSR J0631. The differential upper limits obtained in this work are shown as blue pentagons and red squares in the case of point-like emission and emission from a disk with $r_{pwn} = 0.3°$, respectively. Extrapolating Milagro's estimation to lower energies under the assumption of a power-law with photon index 2.2 (dashed green line), our obtained limits are in tension with their results (Color figure online)

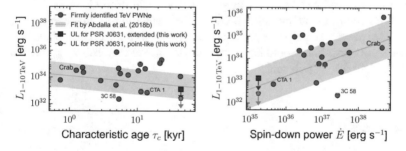

Fig. 9.7 Luminosities $L_{1-10\,\mathrm{TeV}}$ of the pulsar wind nebulae in the 1–10 TeV energy range versus the characteristic age τ_c of the pulsar and its spin-down power $\dot{E}$. We include all 19 firmly identified PWNe from Abdalla et al. [7] and add our upper limit for PSR J0631 in case of a disk-like extension with $r_{pwn} = 0.3°$. The fits provided by Abdalla et al. [7] (shown as straight lines, the shaded regions depict the 1σ variations the luminosities are scattered with) do not only take into account the firm identifications but also upper limits from PWN candidates and non-detections in the H.E.S.S. Galactic Plane Survey [9], and therefore seem to undershoot the data points. The pulsar wind nebulae towards the outer part of our galaxy are labeled. **Left**: $L_{1-10\,\mathrm{TeV}}$ is widely scattered over τ_c and a correlation is statistically unclear. **Right**: The correlation of increasing $L_{1-10\,\mathrm{TeV}}$ with higher spin-down power is statistically stable and suggests a relation of $L_{1-10\,\mathrm{TeV}} \propto \dot{E}^{0.59}$. For both parameters our luminosity limits for PSR J0631 are consistent with the typical scatter of observed values. Plots reproduced from Abdalla et al. [7] (Color figure online)

To put our upper limits in context with the TeV PWNe population of our Galaxy, we convert the integral flux upper limits into luminosity limits $L_{1-10\mathrm{TeV}}$ in the 1–10 TeV energy range assuming a photon index of 2.2 and a distance of 2.1 kpc. Together with the other 19 firmly identified TeV pulsar wind nebulae, we plot the obtained limits versus the characteristic age τ_c and the spin-down power $\dot{E}$ in Fig. 9.7. In their population study [7] report a stable correlation of luminosity with pulsar spin-down with a p-value of 0.010 and suggest a relation of $L_{1-10\,\mathrm{TeV}} \propto \dot{E}^{0.59\pm0.21}$. In contrast, they find the luminosities to be scattered widely over the characteristic ages τ_c and the p-value of 0.13 does not allow to claim a clear statistical correlation. Our results do not pose stringent limits to the fits provided by [7] and are consistent with the typical variations observed in the TeV PWN population so far.

We summarize our findings by emphasizing that our upper limits to a possible pulsar wind nebula around PSR J0631+1036 favor the interpretation of Milagro's hot spot in terms of a statistical artifact and corroborate the non-detection reported by HAWC in [11]. With ∼37 h of MAGIC observations we are able to provide one of the strongest luminosity limits to date for a TeV PWN towards the outer part of our galaxy. A natural explanation for the lack of TeV PWN detections in the outskirts of the Milky Way could be the decreasing interstellar radiation field (ISRF) with increasing distance to the Galactic Center, which would in general diminish the inverse Compton emission from these nebulae. This possibility will be further discussed in a forthcoming paper [12], in which we will have access to all 6 PWN candidates that were observed with MAGIC. Compared to the so far firmly identified TeV PWNe, PSR J0631 seems to exhibit rather border values regarding its characteristic age and spin-down power. Therefore, a possible detection of its pulsar wind nebula by means of deeper observations would have great impact on the established correlations and evolution of these parameters.

References

1. Ahnen M et al (2017a) Performance of the MAGIC telescopes under moonlight. Astropart Phys 94:29–41. https://doi.org/10.1016/j.astropartphys.2017.08.001
2. Aleksić J et al (2015b) The major upgrade of the MAGIC telescopes, part I: the hardware improvements and the commissioning of the system. Astropart Phys, 1–15. https://doi.org/10.1016/j.astropartphys.2015.04.004
3. Zanin R et al (2013) MARS, the MAGIC analysis and reconstruction software. In: 33rd international cosmic ray conference, Rio de Janeiro, Brazil, p 773
4. Nievas-Rosillo M, Contreras JL (2016) Extending the Li&Ma method to include PSF information. Astropart Phys 74:51–57. https://doi.org/10.1016/j.astropartphys.2015.10.001
5. Prandini E et al (2015) Study of hadron and gamma-ray acceptance of the MAGIC telescopes: towards an improved background estimation. In: 34th international cosmic ray conference, The Hague, Netherlands
6. Ahnen M et al (2018) Indirect dark matter searches in the dwarf satellite galaxy Ursa Major II with the MAGIC telescopes. J Cosmol Astropart Phys 2018(03):009–009. https://doi.org/10.1088/1475-7516/2018/03/009
7. Abdalla H et al (2018b) The population of TeV pulsar wind nebulae in the H.E.S.S. galactic plane survey. Astron Astrophys, 612:A2. https://doi.org/10.1051/0004-6361/201629377

8. Li TP, Ma YQ (1983) Analysis methods for results in gamma-ray astronomy. Astrophys J 272:317–324
9. Abdalla H et al (2018a) The H.E.S.S. galactic plane survey. Astron Astrophys 612:A1. https://doi.org/10.1051/0004-6361/201732098
10. Rolke WA et al (2005) Limits and confidence intervals in the presence of nuisance parameters. Nucl Instrum Methods Phys Res Sect A: Accelerators, Spectrometers, Detectors and Associated Equipment 551(2–3):493–503. https://doi.org/10.1016/j.nima.2005.05.068
11. Abeysekara AU et al (2017) The 2HWC HAWC observatory gamma-ray catalog. Astrophys J 843(1):40. https://doi.org/10.3847/1538-4357/aa7556
12. MAGIC Collaboration et al (2018) MAGIC observations on Pulsar Wind Nebulae around high spin-down power Fermi-LAT pulsars. In: preparation

Summary and Conclusions

The studies presented in this thesis cover the most energetic light emitted by pulsars and pulsar wind nebulae. Both source types are known to emit gamma rays in the very-high-energy (VHE) range and have been detected by Imaging Atmospheric Cherenkov Telescopes (IACTs) from the ground. For our studies we mainly used data from the Major Atmospheric Gamma-ray Imaging Cherenkov (MAGIC) telescopes, one of the current three major IACTs in the world. We also analyzed data from the Large Area Telescope (LAT) on board of the *Fermi* satellite, which owing to its wide field of view and overlapping energy range, is perfectly suited to provide observational guidance to IACTs.

The main result of this work is the discovery of pulsed emission from the Crab pulsar up to ~1.2 TeV, the most energetic light ever seen from this class of astrophysical objects. This achievement was rendered possible by an unprecedented large data set of ~414 h from the MAGIC telescopes (see Chap. 5). We were able to detect the inter-pulse P2 in the Crab pulsar's phaseogram at a 5.2σ level above 400 GeV and derived a steep power-law like spectrum with a photon index of 3.13 ± 0.18 in a broad energy range from ~85 GeV to ~1.2 TeV. The main pulse P1, on the other hand, is barely visible at energies above 400 GeV and we could only determine its spectrum up to ~500 GeV. Both peak spectra connect smoothly to the *Fermi*-LAT spectral points above 10 GeV. Joint power-law fits to the MAGIC and *Fermi*-LAT spectral points revealed a significant difference between the photon indexes of the main pulse P1 (3.54 ± 0.09) and the inter-pulse P2 ($3.01 + 0.06$). This sustains the trend of a decreasing intensity ratio $P1/P2$ with energy, as seen by the *Fermi* LAT above 1 GeV, and extrapolates this tendency into the TeV regime.

These results lead to several implications for the emission model of the Crab pulsar. Gamma rays with energies of ~1 TeV must be produced at least ~600 km away from the neutron star to be able to escape the pulsar's magnetosphere, which challenges some of the *slot gap* and *annular gap* models. Regarding the emission mechanism, it seems highly unlikely that pulsed TeV photons are produced via synchro-curvature radiation, which is the standard emission mechanism for gamma-ray pulsars in the GeV regime, since it would require unusual large curvature radii.

© Springer Nature Switzerland AG 2019

D. Carreto Fidalgo, *Revealing the Most Energetic Light from Pulsars and Their Nebulae*, Springer Theses, https://doi.org/10.1007/978-3-030-24194-0

A more plausible mechanism is inverse Compton scattering, which was already put forward after the detection of the >100 GeV pulsation by VERITAS and MAGIC in 2012. Assuming inverse Compton scattering in the Klein-Nishina limit, the Lorentz factors of the accelerated electrons have to be greater than $\sim 2 \times 10^6$ to produce 1 TeV photons. Two models by Aleksić et al. [1, based on the work by Kouichi Hirotani] and [2], among many other models, tried to explain the pulsed emission up to 400 GeV from the Crab pulsar. We discussed both models in the context of the newly found TeV emission and showed that substantial modifications would be necessary in the case of Aharonian et al. [2]. Open questions that have yet to be addressed by theoretical models are the narrow peaks observed in the pulse profile above 100 GeV, the decreasing intensity ratio $P1/P2$ with energy and the phase coherence of the peaks along the entire electromagnetical spectrum, from radio up to TeV energies.

The observation of pulsed TeV photons provides a unique set of data to investigate fundamental physics. In Chap. 6 we used our large data set of the Crab pulsar to test for Lorentz Invariance Violation (LIV) in terms of a wavelength dependent speed of light. For this purpose we modeled the gamma-ray emission of the inter-pulse incorporating an energy dependent (linearly or quadratically) group velocity of photons for a *subluminal* and *superluminal* scenario. By applying Bayesian inference to obtain our model parameters, we were able to derive conservative 95% lower limits on the invariant energy scales, assuming either a linear or quadratic energy dependence. Our quadratic limits $E_{QG_2} > 3.1 \times 10^{10}$ GeV and $E_{QG_2} > 1.6 \times 10^{10}$ GeV, for a subluminal and superluminal scenario, respectively, corroborate the principle findings by Ahnen et al. [3] and are a factor of $\lesssim 6$ worse than the current world-best limits. They show, however, much room for improvement considering the upcoming Cherenkov Telescope Array (CTA) observatory.

While over 200 pulsars are known to emit gamma rays in the MeV to GeV regime, only two pulsars have been detected in the VHE range to date: the Crab and the Vela pulsar. Against this background, in Chap. 7 we conducted a search for a possible next very-high-energy pulsar looking through the wealth of gamma-ray pulsars discovered by *Fermi*-LAT. Since we wanted to make use of the pulsars' timing information in our search, we restricted ourselves to millisecond pulsars (MSPs), which are easier to time compared to canonical pulsars. We focused on MSPs that would be well observable from the MAGIC site, and further included Geminga and PSR J0614- 3329 to our sample, two known and promising candidates for VHE emission. Using almost 90 months of *Fermi*-LAT data, we found evidence for >50 GeV pulsation from the MSP PSR J0614- 3329, which seems to be a compelling candidate for follow-up observations by IACTs in the Southern Hemisphere.

In Part III of this thesis we put aside VHE emission from pulsars and focused on TeV emission from pulsar wind nebulae (PWNe). Motivated by a Milagro hot spot (3.7σ) at ~ 35 TeV, we conducted observations in the direction of the young and energetic gamma-ray pulsar PSR J0631+1036 with the MAGIC telescopes to look for steady TeV emission from a possibly extended PWN around the pulsar. We collected a total amount of ~ 37 h of high quality data, but found no hint of a signal in our analysis. 95% confidence upper limits on the integral flux F_{300} above

300 GeV were computed assuming a photon index of 2.2 and under the assumption of a point-like emission ($F_{300} < 6.0 \times 10^{-13}$ cm^{-2} s^{-1}) and a disk-like emission profile with radius $r_{\mathrm{pwn}} = 0.3°$ ($F_{300} < 2.8 \times 10^{-12}$ cm^{-2} s^{-1}). These limits are in tension with Milagro's results and corroborate the non-detection recently reported by HAWC. Putting the limits in the context with the TeV PWN population study by Abdalla et al. [4], we concluded that our luminosity limits for a possible TeV PWN around PSR J0631 lie within the observed scatter of TeV PWN luminosities in our galaxy.

During the course of his Ph.D. studies, the author of this thesis spent several months at the Roque de los Muchachos Observatory in La Palma, Canary Islands (Spain), to carry out observations for the MAGIC collaboration. He also participated in the refinement of the analysis of data taken with a new trigger system and made significant improvements to the on-site analysis (OSA) chain, for which he was one of the persons in charge and which provides the low-level analysis products to the whole collaboration. His contributions supported the scientific output by the MAGIC collaboration, whose results have always made and continue to make a significant impact in the field of VHE astrophysics.

Regarding the future of VHE gamma-ray astronomy, a key player for its progress will be the upcoming Cherenkov Telescope Array (CTA) Observatory. This major project is ran by over 1400 members in 32 countries,[1] and will consist of more than 100 telescopes located in the Northern and Southern Hemisphere providing a nearly full-sky coverage. CTA's sensitivity will outperform current generation IACTs by an order of magnitude and most likely improve many of the results presented in this thesis. With far less observation time, CTA will be able to either measure the Crab pulsar spectrum up to energies far beyond $\sim$1 TeV or detect a cut-off in its VHE tail. This should allow us to stringently limit the LIV energy scales and should provide world-best limits for the quadratic term. Furthermore, the low energy threshold of CTA's Large Size Telescopes (LSTs) of about 25 GeV will presumably increase the number of gamma-ray pulsars detected by IACTs, and thus contribute to a better understanding of the scarce VHE component in pulsars. Concerning pulsar wind nebulae, one of CTA's Key Science Project is the survey of the full Galactic plane. This survey will be a factor of 5–20 more sensitive than current attempts and is expected to add many more members to the population of TeV PWNe, not only towards the Galactic Center but also from the outer part of the galaxy [5]. All in all, CTA seems a promising tool for studying the most energetic light from pulsars and their nebulae.

---

[1] https://www.cta-observatory.org, last accessed 11/04/2018.

References

1. Aleksić J et al (2012) Phase-resolved energy spectra of the Crab pulsar in the range of 50–400 GeV measured with the MAGIC telescopes. Astron Astrophys 540:A69. https://doi.org/10.1051/0004-6361/201118166
2. Aharonian FA et al (2012) Abrupt acceleration of a 'cold' ultrarelativistic wind from the Crab pulsar. https://doi.org/10.1038/nature10793
3. Ahnen ML et al (2017) Constraining Lorentz invariance violation using the Crab pulsar emission observed up to TeV energies by MAGIC. Astrophys J Suppl Ser 232(1):9. https://doi.org/10.3847/1538-4365/aa8404
4. Abdalla H et al (2018) The population of TeV pulsar wind nebulae in the H.E.S.S. galactic plane survey. Astron Astrophys 612:A2. https://doi.org/10.1051/0004-6361/201629377
5. The Cherenkov Telescope Array Consortium, et al (2017) Science with the Cherenkov Telescope Array

Appendix A
Interaction Processes in Very-High-Energy Astrophysics

Cosmic gamma rays are primarily produced in interactions of charged, energetic cosmic rays (electrons/positrons and protons) with ambient electromagnetic fields or matter. Gamma rays are therefore able to map densities and energetics of cosmic rays as well as their interaction partners, often referred to as "targets". Here we will superficially sketch the most relevant interaction processes for this theses, including the ones relevant for the instrumental detection of very-high-energy (VHE) gamma rays. Detailed treatments can be found in the canonical publications and textbooks by Blumenthal and Gould [1], Rybicki and Lightman [2], Jackson [3], Aharonian [4], Longair [5].

Bremsstrahlung   Also referred to as *free-free emission*, Bremsstrahlung is the radiation of an unbound charged particle due to its acceleration in the Coulomb field of another charged particle. The intensity spectrum radiated by a single electron in the field of a nucleus is constant up to a cut-off frequency, at which it decays exponentially. Bremsstrahlung is the dominant energy-loss mechanism for electrons and positrons entering the atmosphere. In the relativistic limit the energy loss rate $-dE/dt$ is proportional to E, resulting in the exponential loss of energy by the electron. It is common to define a radiation length X_0 for which the electron looses a fraction $(1 - 1/e)$ of its energy. In astrophysics the bremsstrahlung of a thermal plasma is called *thermal bremsstrahlung* and is primarily observed in the X-ray band from hot intracluster gas in a galaxy cluster.

Synchrotron radiation   The motion of a charged particle in a uniform static magnetic field B will generally consist of a constant velocity along the magnetic field lines and a circular motion about it, resulting in a spiral path with constant pitch angle α (see Fig. A.1, left panel). In the non-relativistic case ($v \ll c$), when the beaming of the radiation can be neglected, an electron (charge e) with the mass m_e emits energy at the non-relativistic gyrofrequency $\nu_g = eB/2\pi m_e$ or $\nu_g = 28\,\mathrm{GHz\,T^{-1}}$. In the ultrarelativistic limit ($v/c \approx 1$, $\gamma \gg 1$, where v is the velocity of the charged particle, c the speed of light and γ denotes the Lorentz factor), the Doppler and aberration effects result in a spread of emitted frequencies and the radiation may be regarded as a continuous spectrum. The spectral photon

© Springer Nature Switzerland AG 2019
D. Carreto Fidalgo, *Revealing the Most Energetic Light from Pulsars and Their Nebulae*, Springer Theses, https://doi.org/10.1007/978-3-030-24194-0

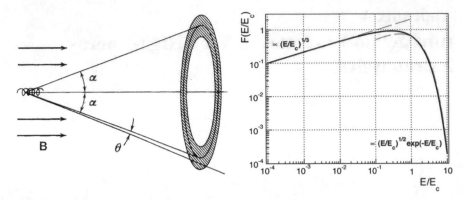

Fig. A.1 Characteristics of synchrotron radiation. **Left**: Trajectory of a charged particle in a uniform static magnetic field with pitch angle α. In the ultrarelativistic limit the radiation is confined to the shaded solid angle ($\theta \propto 1/\gamma$) in the direction of the electron velocity, forming a cone with half-angle α. Figure adopted from [5]. **Right**: The Function $F(x)$, see Eq. A.2, plotted in logarithmic space. E is the energy of the emitted photon by the synchrotron process and E_c a defined critical Energy (see Eq. A.3)

flux per unit time $N(E)$ of a single electron can be written as

$$N(E) = \frac{\sqrt{3}e^3 B \sin \alpha}{4\pi \varepsilon_0 c m_e h} F(E/E_c), \tag{A.1}$$

$$F(x) = x \int_x^\infty K_{5/3}(z)\mathrm{d}z, \tag{A.2}$$

$$E_c = \frac{3}{2}\gamma^2 h \nu_L \sin \alpha, \tag{A.3}$$

where ε_0 is the electric constant, h the Planck constant and $K_{5/3}$ is the modified Bessel function of order $5/3$. E_c is the so-called *critical energy* at which the emission spectrum roughly reaches his maximum ($E_{max} \approx 0.29E_c$). The shape of the spectrum is defined by the function $F(x)$ and is shown in the right panel of Fig. A.1. The asymptotic expressions for function $F(x)$ are

$$F(x) \propto \begin{cases} (E/E_c)^{1/3} & \text{for } E \ll E_c \\ (E/E_c)^{1/2} \exp(-E/E_c) & \text{for } E \gg E_c, \end{cases} \tag{A.4}$$

indicating an exponential cutoff at energies above E_c. The total energy loss rate is proportional to

$$-\left(\frac{\mathrm{d}E}{\mathrm{d}t}\right) \propto B^2 \gamma^2 \sin^2 \alpha. \tag{A.5}$$

The time resulting from the division of the electron energy ($E_e = \gamma c m_e^2$) by the total energy loss rate is also known as the *synchrotron cooling time*.

Assuming that the electron energies exhibit a power-law distribution with spectral index p, the spectrum of the synchrotron radiation also follows a power-law with the index $(p - 1)/2$ [5].

Curvature radiation In a curved magnetic field, assuming that its curvature is much bigger than the gyroradius, a charged particle will move along the bended magnetic filed lines. Since it will be accelerated transversely, in addition to the synchrotron radiation it will radiate another component called *curvature radiation*. By presuming that the curved trajectory of the particle is due to an introduced virtual magnetic field, and not because of the curved magnetic field lines, one can easily establish an analogy to the closely related synchrotron radiation. In the ultrarelativistic case and adopting a pitch angle of $\sin \alpha = 1$, the gyroradius of an electron is written as $r = \gamma m_e c / eB$. Adopting the analogy illustrated in Fig. A.2, we introduce the virtual magnetic field

$$B_{\text{curv}} = \frac{\gamma m_e c}{r_{\text{curv}} e}, \tag{A.6}$$

r_{curv} being the radius of the curved magnetic field line. Replacing the magnetic field in Eqs. A.1 and A.3 with B_{curv} one obtains the corresponding quantities for the curvature radiation. The critical energy and the energy loss rate in case of curvature radiation are

$$E_{c,\text{curv}} = \frac{3}{2} \hbar c \gamma^3 \frac{1}{r_{\text{curv}}} \quad \text{and} \tag{A.7}$$

$$-\left(\frac{dE}{dt}\right)_{\text{curv}} \propto \gamma^4 / r_{\text{curv}}^2. \tag{A.8}$$

The maximum Lorentz factor γ_{max} of an electron emitting curvature radiation, which is continuously being accelerated by an electric field E, is limited by the equilibrium of energy loss rate (see Eq. A.8) and energy gain rate ($dE/dt = eEc$),

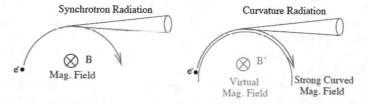

Fig. A.2 Sketch of the analogy between synchrotron and curvature radiation. **Left**: Trajectory for an electron in a magnetic field emitting synchrotron radiation. **Right**: Electron moving along a strong magnetic field line. The trajectory of the electron can be explained by introducing a virtual magnetic field. Figure taken from Saito [6]

$$\gamma_{max} \propto E^{\frac{1}{4}} \sqrt{r_{curv}} .$$ (A.9)

Assuming a power-law spectrum for the electrons' energies with spectral index p, the spectrum of the curvature radiation follows a power-law with the index $(p-2)/3$ [7].

In the most general case, electrons can be accelerated along a curved magnetic field line and spiral around them at the same time. This leads to the formulation of a *synchro-curvature radiation*, for which a detailed summary can be found in the excellent review by Vigano et al. [8]. They introduce a synchro-curvature parameter $\xi = (r_{curv} \sin^2 \alpha)/(r_{gyr} \cos^2 \alpha)$, where r_{gyr} is the relativistic gyroradius. For $\xi \ll 1$, the curvature radiation dominates the emission, if $\xi \gg 1$, synchrotron losses start to dominate.

Inverse Compton scattering   In Inverse Compton (IC) scattering, the ultrarelativistic electron scatters a low energy photon ε to higher energies ε' so that the photon gains energy at the expense of the kinetic energy of the electron. The maximum energy transfer occurs for head-on collisions between the electron and the photon. In this scenario we can write the photon energy after the scattering as (see Eq. 2.52 in [1])

$$\varepsilon' = \frac{4\varepsilon\gamma^2 m_e c^2}{m_e c^2 + 4\varepsilon\gamma},$$ (A.10)

where m_e is the rest mass of the electron and γ its Lorentz factor. In the limit of $\gamma\varepsilon \ll m_e c^2$ (also referred to as *Thompson limit*) the maximum scattered photon energy can be written as $\varepsilon' \sim 4\varepsilon\gamma^2$, on the other hand in the *Klein-Nishina limit*, that is $\gamma\varepsilon \gtrsim m_e c^2$, the maximum photon energy is $\varepsilon' \sim \gamma m_e c^2$. In the Thompson limit the electron tends to loose small fractions of its energy in a continuous series of Compton scatterings, while in the Klein-Nishina regime the scattered photon carries away a large part of the electron's energy in a single scattering process. A power-law spectrum of the electrons' energies with spectral index p, results in a power-law energy spectrum of the scattered radiation with index $(p-1)/2$ [5].

The case in which electrons upscatter low energy photons emitted by themselves through synchrotron radiation, is known as *synchrotron-self-Compton* (SCC) radiation.

π^0 **decay**   The most important gamma-ray emission process by hadrons is the decay of a neutral pion into two gamma rays. Neutral pions are mostly produced in inelastic collisions between hadronic cosmic rays (protons or α particles) and have a decay lifetime of only 8.4×10^{-17} s after which they enter the decay channel of $\pi^0 \rightarrow \gamma + \gamma$ with a probability of ~ 0.99.

The produced high-energy gamma-ray spectrum follows the energy spectrum of the pions and essentially reproduces the spectrum of the parent protons. Very energetic hadronic collisions also produce charged pions with comparable probabilities, which decay into neutrinos. In astrophysics the detection of correlated neutrino and gamma-ray fluxes is therefore a strong indication for hadronic acceleration mechanisms at the emission site.

Pair production For very-high-energy (VHE) photons the pair production process dominates the photon cross section when interacting with matter [9]. Above the energy threshold of $2m_ec^2$, m_e being the rest mass of an electron and c the speed of light, a photon is able to decay into an electron-positron pair ($e^{\pm}$) in the field of a nucleus or an electron ($\gamma \rightarrow e^+ + e^-$). The field is necessary, since one-photon pair production cannot conserve both energy and momentum in field-free space. The cross sections of Bremsstrahlung and pair production are closely related through their similar Feynman diagrams. Therefore, the mean free path of pair production X_p can be expressed by the radiation length of Bremsstrahlung X_0 via $X_p = (9/7)X_0$.

In pulsars the strong magnetic field is able to consume the extra momentum of one photon pair production and the resulting *magnetic pair creation* is an important absorption mechanism that prevents the escape of gamma rays from the pulsar's magnetosphere. In contrast to one-photon pair production, the mechanism of two-photon pair production also works in a field-free environment. In astrophysics this mechanism explains the attenuation of very-high-energy gamma rays from far away sources through their interaction with the extragalactic background light (EBL, $\gamma_{VHE} + \gamma_{EBL} \rightarrow e^+ + e^-$).

Photon splitting One photon can divide in two or more photons with lower energies in the presence of a strong magnetic field. In pulsars the rate of this process is generally much smaller than that of magnetic pair creation above the energy threshold $\varepsilon = 2m_ec^2/\sin\theta$, where θ is the angle between the magnetic field and the photon's velocity vector [10]. Below the threshold ε, however, it significantly contributes to the absorption of energetic photons from neutron stars magnetospheres.

References

1. Blumenthal GR, Gould RJ (1970) Bremsstrahlung, synchrotron radiation, and compton scattering of high-energy electrons traversing dilute gases. Rev Mod Phys 42(2):237–270. https://doi.org/10.1103/RevModPhys.42.237
2. Rybicki GB, Lightman AP (1979) Radiative processes in astrophysics. Wiley-Interscience, New York (ISBN 0471827592)
3. Jackson JD (1998) Classical electrodynamics, 3rd edn. Wiley, New York (ISBN 047130932X)
4. Aharonian FA (2004) Very high energy cosmic gamma radiation: a crucial window on the extreme universe. World Scientific, Singapore (ISBN 9812561730)
5. Longair MS (2011) High energy astrophysics, 3rd edn. Cambridge University Press, Cambridge (ISBN 0521756189)
6. Saito T (2010) Study of the high energy gamma-ray emission from the Crab pulsar with the MAGIC telescope and Fermi-LAT. PhD thesis, LMU
7. Zhang B, Dai ZG (2011) Synchro-curvature self-compton radiation of electrons in curved magnetic fields. Mon Not R Astron Soc 414(4):2785–2792. https://doi.org/10.1111/j.1365-2966.2010.18187.x
8. Vigano D et al (2014) Compact formulae, dynamics and radiation of charged particles under synchro-curvature losses. Mon Not R Astron Soc 447(2):1164–1172. https://doi.org/10.1093/mnras/stu2456

9. Groom DE, Klein SR (2000) Passage of particles through matter. Euro Phys J C 15(1–4):163–173. https://doi.org/10.1007/BF02683419
10. Harding AK, Lai D (2006) Physics of strongly magnetized neutron stars. Rep Prog Phys 69(9):2631–2708. https://doi.org/10.1088/0034-4885/69/9/R03

Appendix B
Pulsar Timing

When analyzing pulsar data one has to take into account that the times of arrival (TOAs) of the pulses at the observatory on Earth, do not directly reflect the timing of the pulsed emission at the pulsar site. A general relativistic frame transformation between observatory proper time and pulsar proper time is necessary to compute the *pulse phase* of the emission via a model of the intrinsic variations in the pulse period. Pulse phases are usually given in the range of [0, 1[or [0, 360[corresponding to a full rotation of the neutron star.[2] In the following we will roughly sketch out how to compute pulse phases. For an comprehensive introduction to pulsar timing the reader is referred to the excellent descriptions by [1, 2].

Inertial Reference Frame

The center of mass of the solar system, known as Solar system barycentre (SSB), moves essentially uniformly through space and is therefore a convenient inertial frame of reference. It is common to transform the observed TOA, t_{obs}, to an equivalent TOA for the same pulse wave-front at the SSB, which requires the quantitative calculation of a numerous of time delays originating from geometrical, relativistic and dispersion effects. The propagation delay from the pulsar to the SSB is usually ignored, but orbital motions of the pulsar, if present, must indeed be taken into account. Considering the most significant ones, we obtain following expression for the time of emission (for a complete discussion of all the time delays, the reader is referred to [2]):

$$t_e = t_{obs} - \triangle_C - \triangle_{R\odot} - \triangle_{E\odot} - \triangle_{S\odot} - \triangle_D - \triangle_B, \qquad (B.1)$$

---

[2]It is common practice, however, to show two rotations in the pulsar light curves for clarity.

© Springer Nature Switzerland AG 2019
D. Carreto Fidalgo, *Revealing the Most Energetic Light from Pulsars and Their Nebulae*, Springer Theses, https://doi.org/10.1007/978-3-030-24194-0

where (typical values for the delays are given in brackets, taken from Hobbs et al. [1]):

- $\triangle_C$ [∼1 µs]: The *Clock delay* takes into account that the TOAs are measured against a local clock at the observatory. Usually these measurements are transformed to a relativistic dynamical time scale such as TDB (Barycentric Dynamical Time) or TCB (Barycentric Coordinate Time).
- $\triangle_{R\odot}$ [∼500 s]: The *Roemer delay* is the vacuum delay between the arrival of the pulse at the observatory and the Solar system barycenter. Normally the distance from the observatory to the pulsar is much bigger than the distance to the SSB ($|\mathbf{r}_{op}| \gg |\mathbf{r}_0|$, see Fig. B.1) and the curvature of the wave-front connecting photons simultaneously emitted from the pulsar can be neglected.
- $\triangle_{E\odot}$ [∼1.6 ms]: The *Einstein delay* is due to the relativistic space-time transformation of the coordinate frame of the observatory to the quasi-inertial frame of the Solar system barycenter. While the relativistic length contraction is negligible, the time dilation is not.
- $\triangle_{S\odot}$ [∼112 µs]: The *Shapiro delay* accounts for the time delay caused by the passage of the pulse through curved space-time and is obtained considering all the bodies in the Solar system.
- $\triangle_D$: The dispersion delay affects primarily radio signals that encounter significant dispersion in the interplanetary (Solar wind) as well as in the interstellar medium. Since this delay is inversely proportional to the frequency of the observed photon ($\triangle_{D\odot} \propto \nu^{-2}$), in the case of gamma rays this term is neglected.
- $\triangle_B$: Especially millisecond pulsars are often found in binary systems. This term takes into account any orbital motion of the pulsar as well as further effects of the companion, such as an Einstein delay and a Shapiro delay.

Modeling the Pulse Phase

After relating a measured TOA to a time of emission, one now can try to model the intrinsic variations in the pulse period. A basic characteristic of a pulsar is the precise period P modulated by a slow increase due to a gradual loss of rotational energy ($dP/dt \simeq 10^{-15}$). In most cases a Taylor expansion with two or three terms is sufficient to predict accurately enough the pulse phase ϕ (see [2]):

$$\phi(t) = \sum_{n \geq 1} \frac{\nu^{n-1}}{n!} (t_e - t_P)^n + \phi_0, \qquad (B.2)$$

where $[\nu^{n-1}]$ are the frequency derivative terms and our fit parameters, while t_P is the epoch in which $\dot{\phi} = \nu$ and is set by the user. ϕ_0 is introduced to achieve absolute phase alignment and is normally defined in terms of a reference TOA for a specific observing site and frequency. The fractional part of ϕ is then used to assign pulse

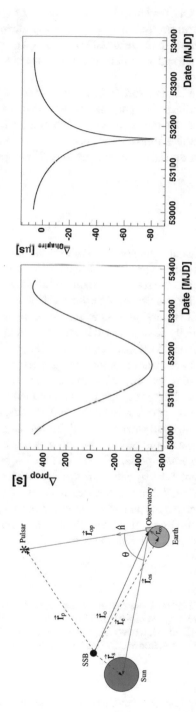

Fig. B.1 Left: Sketch of the *Roemer delay*. $|\mathbf{r}_{op}|$ is assumed to be much bigger than $|\mathbf{r}_0|$. **Middle:** The Roemer delay as a function of time. The delay is plotted over a time span of one year. **Right:** The *Shapiro delay* as a function of time. Figures adopted from [3]

phases to observed events. One way to fit Eq. B.2 to our TOAS $t_{e,i}$ is to define a priori a pulse phase model ϕ_a with an approximate frequency (by means of, for example, the Fourier transform) and minimize the residuals $\phi_a(t_{e,i}) - N_i$, where N_i is the nearest integer to $\phi_a(t_{e,i})$. This is normally an iterative procedure in which ϕ_a gets updated with the post-fit parameters.

The set of parameters in Eq. B.2 form a basic *ephemeris* of the pulsar, which can be arbitrarily expanded with further parameters reflecting a more complex timing model. With a given pulsar ephemeris we can assign pulse phases to our events and *phase fold* them to obtain pulsar light curves, also referred to as *phaseograms*. Phase folding consists of binning the data with respect to their assigned pulse phases.

Timing Irregularities

Although pulsars show a remarkable precision in its rotation, there are random irregularities in the periods that are measured as phase deviations in the pulses' TOAs. This *timing noise* is often given as the root-mean-square (RMS) of the residuals between the measured and predicted TOAs. Especially young pulsars exhibit large timing noise of up to ~1 s, whereas millisecond pulsars rotate far more stable (~1 μs, see [4]). Pulsar timing noise can usually be subdivided into a *white* component (equal power at all fluctuation frequencies) and a *red* component (greater power at lower fluctuation frequencies). A proper characterization of the timing noise can mitigate its affect on other parameters of the timing model, such as the timing position of the pulsar, and is therefore desirable when aiming for high precision ephemerides [5].

In contrast to the continuously erratic behavior of the timing noise, pulsars can also show pronounced step changes in rotation speed, known as *glitches*. These events, resulting in a short decrease of the pulsar's period, are rare and one third of those observed took place in the young pulsars Crab and Vela.[3] A possible origin of these sudden spin-ups are thought to be so-called *starquakes*, where the slow down of the pulsar decreases the centrifugal force on the stellar's surface and an abrupt crack of the neutron star's crust changes the pulsar's moment of inertia [6]. While on average a pulsar with a characteristic time τ_c [kyr] will glitch a maximum of $(6 \pm 2) \times \tau_c^{-0.48}$ times per year [7], the Crab pulsar shows an average glitch rate of about ~1 per year.

References

1. Hobbs GB et al (2006) TEMPO2, a new pulsar-timing package - I. An overview. Mon Not R Astron Soc 369(2):655–672. https://doi.org/10.1111/j.1365-2966.2006.10302.x
2. Edwards RT et al (2006) TEMPO2, a new pulsar timing package - II. The timing model and precision estimates. Mon Not R Astron Soc 372(4):1549–1574. https://doi.org/10.1111/j.1365-2966.2006.10870.x

---

[3] A table of observed glitches from the Crab Pulsar can be found at: http://www.jb.man.ac.uk/pulsar/glitches/gTable.html, last accessed 07/03/2018.

3. López MM (2006) Astronomíia Gamma con el Telescopio MAGIC: Observaciones de la Nebulosa y Pulsar del Cangrejo. PhD thesis
4. Shannon RM, Cordes JM (2010) Assessing the role of spin noise in the precision timing of millisecond pulsars. Astrophys J 725(2):1607–1619. https://doi.org/10.1088/0004-637X/725/2/1607
5. Kerr M et al (2015) Timing gamma-ray pulsars with the fermi large area telescope: timing noise and astronometry. Astrophys J 814(2):128. https://doi.org/10.1088/0004-637X/814/2/128
6. Lattimer JM, Prakash M (2004) The physics of neutron stars. Science 304(5670):536–542. https://doi.org/10.1126/science.1090720
7. Espinoza CM et al (2011) A study of 315 glitches in the rotation of 102 pulsars. Mon Not R Astron Soc 414(2):1679–1704. https://doi.org/10.1111/j.1365-2966.2011.18503.x

Appendix C
The On-Site Analysis Chain of MAGIC

This appendix describes an up-to-date status of the *On-Site Analysis* (OSA) chain operating at the MAGIC site (see Sect. 3.3) and was presented at the ICRC 2015 in form of a poster contribution [1]. As one of the persons in charge of OSA, the author of this thesis continuously worked on it during the course of his Ph.D. studies improving and maintaining its workflow.

The fast processing of the data at the observation site plays an essential part in the operation of the telescopes. OSA provides provisional intermediate-level analysis products that allow for a fast offline analysis, and thus a quick assessment of the nights data in the case of targets of opportunity like flaring sources. Moreover, the members of the MAGIC collaboration normally use the low-level analysis products provided by OSA for their scientific analyses.

OSA has steadily improved since the beginning of the experiment and we present here the status of the system, including the latest upgrades and details on its performance. The first section provides an overview of the computing infrastructure available at the MAGIC site and the organization of the data flow across the different systems. The second section is devoted to OSA's pipeline, its workflow and the performance.

Computing Infrastructure and Data Flow

The computing system at the MAGIC site consists of a cluster of computers linked by internal networks and accessible from outside via a public web server.[4] Data Acquisiton (DAQ) and operations are carried out in the DAQ and subsystem servers, while data processing takes place in the analysis cluster. The later is composed of a set of high performance computing (HPC) servers running the same operating system (Scientific Linux Cern 6.3) and sharing both network and disk access. The storage capacity of the whole cluster is provided by four RAID systems and local disks

---

[4]http://www.magic.iac.es/, last accessed 19/04/2018.

© Springer Nature Switzerland AG 2019
D. Carreto Fidalgo, *Revealing the Most Energetic Light from Pulsars and Their Nebulae*, Springer Theses, https://doi.org/10.1007/978-3-030-24194-0

for temporary storage. Each RAID system has a different capacity, configuration and purpose. Volumes devoted to DAQ (RAID1 and RAID2 for the two MAGIC telescopes M1 and M2, respectively), use the XFS file system and are handled by the DAQ machines. Shared volumes (RAID3 and RAID4) store the compressed raw data, software and user data. They are connected by means of a fibre-channel dedicated network, and use a GFS2-formatted file system accessible to every machine of the cluster. The decoupling of the two storage arrays is important since it allows the on-site activities (like the on-site analysis and the data check) to act independently from the data acquisition and to maintain legacy systems used by the DAQ machines isolated from the more modern analysis cluster. Data created at the telescopes ($\sim$100 GB per telescope per hour of observation) are processed in a well-defined chain, comprising different activities according to the logical actions and subsystems involved. The responsibility for each activity relies in a different institute belonging to the MAGIC collaboration, which takes care of the full development and deployment of the services and tools required for a correct processing. The data flow of the activities carried out on-site is sketched in Fig. C.1.

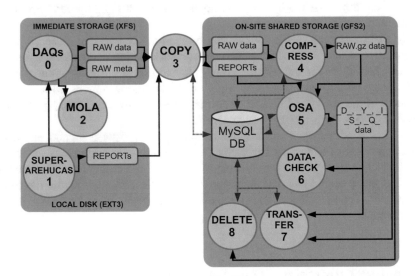

Fig. C.1 Data flow scheme at the MAGIC site. SuperArehucas (1) is the central control for the MAGIC telescopes, controlling the data taking and merging the information of all important subsystems into report files. The copy/compress activities (3, 4) are done simultaneously, while the others follow approximately the sequential order indicated by the number inside the circle. The output of the on-site analysis (5) are low-level analysis products with different key names (*_D_*, *_Y_* etc.). The transfer of the data to the MAGIC data center (7) and the deletion (8) are triggered by the MySQL database that keeps track of the daily activities on-site. For more details on the two processes MOLA and DataCheck the reader is referred to [1, 2]. Image courtesy of Alejandro Lorca

The On-Site Analysis

The aim of OSA is to provide low and intermediate level analysis products to the collaboration on the day after the observation night. OSA essentially takes care of the calibration of the data, the cleaning and parameterization of the images and merges the data from the two MAGIC telescopes. Those steps reduce the file sizes by a factor of ~200 compared to the compressed raw data and are performed in parallel with the raw data transfer to the MAGIC data center located at the *Port d'Informació Científica* (PIC) in Barcelona. OSA consists of the MAGIC *Analysis and Reconstruction Software* (MARS, [3]), a set of python scripts, Unix cron jobs and a PBS/Torque resource manager to allow for a high degree of parallelization using the 40 cores assigned to OSA in the computer cluster at the MAGIC site. Its objective is achieved by a high degree of parallelization when processing the data and by starting its operation already during the observation night.

OSA and MARS

MARS is a set of C++ classes based on the well known ROOT package from CERN[5] and allows the analysis of MAGIC data by means of compiled programs called MARS executables. OSA interacts with those executables via sub-processes, which are spawned within the python scripts, passing them arguments and evaluating their return codes. The main MARS executables used by OSA are the following:

- *sorcerer*: takes as input the compressed raw data and calibrates it.
- *merpp*: adds the central control reports to the calibrated files.
- *star*: performs the cleaning and the parametrization of the images.
- *superstar*: combines the M1 and M2 *star* files, containing image parameters, and performs the stereo reconstruction.
- *melibea*: estimates the event properties, for example energy, direction and particle kind.

In general, collaboration members start their analyses with *star* or *superstar* files provided by OSA since the next step (*melibea*) requires input generated with the help of Monte Carlo simulations, which on the other hand depend on the instrument setup and analysis goals. Another reason is that OSA does not make any kind of data selection regarding the quality of the data. This is usually done by the analyzer at the *star* or *superstar* level. *Melibea* files are produced to allow for a quick but preliminary off-line analysis of targets of opportunity.

---

[5]https://root.cern.ch, last accessed 28/02/2018.

OSA Workflow

Every hour, the copy/compress process (see Fig. C.1) looks for new raw data on the immediate (DAQ) storage and copies it together with the corresponding report files to the on-site shared storage, classifying it by observation nights. With the same frequency and after a short delay, the on-site analysis chain is initiated by cron jobs that execute the so-called *sequencer* (see Fig. C.2). This script, with the help of the *nightsummary* script, checks for available raw data in the shared storage system and gathers information from the report files for a given observation night. The *sequencer* then goes on to create a list of sequences to be analyzed, establishes relationships between them, making the execution of some of them dependent on the successful completion of others, and sends them to the Torque queue system. Three types of sequences are built, depending on the content of the input files:

- *Calibration*: to process the calibration runs taken for each source before starting the observation.
- *Data*: to process a data run of one telescope, which by default have an observation time of 20 min.
- *Stereo*: to merge and process the output files of the *Data* sequences for each telescope sharing the same data run.

While the command line arguments M1 and M2 build *Calibration* and *Data* sequences for the corresponding telescope, ST results in a list of *Stereo* sequences. The sequences are processed in the working nodes of the computer cluster and handled by different scripts according to their type (*calibrationsequence*, *littlesequence* and *stereosequence*, respectively). The Torque system balances the charge among the nodes.

The standard output of the *sequencer* is a human readable table displaying the characteristics, status and progress of each sequence. This table is automatically

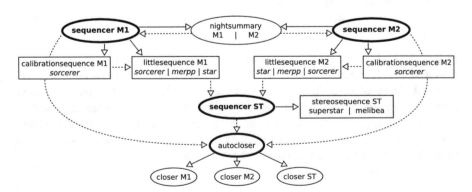

Fig. C.2 A sketch of the OSA workflow. Ellipses indicate scripts run in the user interface (bold ones are initiated be cron jobs) while scripts inside boxes are executed anywhere in one of the working nodes. Solid lines represent job submissions or process calls while dashed lines illustrate dependencies. The MARS executables called by the respective python scripts are written in italic

interpreted by the *autocloser* script that is executed by a cron job on the half-hour. When it detects that the analysis has finished for all the sequences of M1, M2 or ST, the *autocloser* calls the *closer* script. This *closes the day* for the corresponding data (M1, M2, or ST), triggering the transfer of the respective analysis products to the MAGIC data center through an entry in the MySQL database. The output of the sequencer is also copied periodically to a web page for visual inspection by the OSA team.

OSA Performance

The first version of OSA was installed in 2005 and consisted of a small set of shell scripts called by cron jobs. Since then it underwent several updates/changes with a significant remodeling during 2012 when the whole system was transferred to python and the storage system was separated into the immediate storage and the on-site shared storage system. This decoupling allows OSA to start the analysis chain already during the night after the first data is available without disturbing the rest of the data taking.

At the end of 2014, a change in the *nightsummary* script and some helper modules permitted the automatic analysis of some non-standard observations. Additionally, the introduction of the *autocloser* script, developed by the author of this thesis, automatized the interpretation of the *sequencer* table and the *closing of the day*, tasks that formerly had to be taken care of manually. Both modifications led to a higher degree of automatization and significantly decreased the workload for the OSA staff.

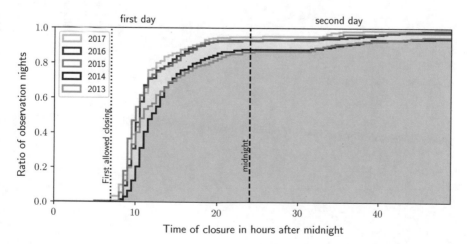

Fig. C.3 OSA performance for closing the stereo analysis (ST) for the last 5 years of operation. The y-axis shows the ratio of nights closed before × hours after midnight. Statistics could only be gathered from 2013 on. Image courtesy of Mireia Nievas-Rosillo (Color figure online)

The performance of OSA over the last 5 years measured by the time needed to process the data of a single day is shown in Fig. C.3. As for 2017, OSA completes the analysis of M1 and M2 for ~85% of the observation nights within the first 12 h after midnight, while for the stereo analysis it is ~80%. About 95% of the observation nights are completely analyzed and closed within the first 24 h. For the rest of the nights, OSA encounters major problems that require a deeper intervention by the OSA staff and are normally fixed within three days.

Together with a stable transfer of the analysis products to the MAGIC data center, where it becomes accessible to the general collaboration, OSA provides the possibility of a fast offline analysis and therefore allows for an adaptive scheduling of targets of opportunity like flaring sources.

References

1. Fidalgo D et al (2015) Data processing activities at the MAGIC site. In: 34th international cosmic ray conference, The Hague, Netherlands, p 8
2. Tescaro D et al (2013) The MAGIC telescopes DAQ software and the on-the-fly online analysis client. In: 33rd international cosmic ray conference, Rio de Janeiro, Brazil, pp 2–5
3. Zanin R et al (2013) MARS, the MAGIC analysis and reconstruction software. In: 33rd international cosmic ray conference, Rio de Janeiro, Brazil, p id 773

Appendix D
Details on the Crab Pulsar Analysis

This appendix provides some further details on the analysis of the Crab pulsar presented in Chap. 5. The individual sections correspond to references made in the main text of Chap. 5.

Cleaning Methods and Cleaning Levels

The image cleaning parameters reflect to a certain extent the hardware configuration of the telescopes at the time (see Sect. 3.3 for an overview of the hardware changes). Table D.1 summarizes the image cleaning methods and cleaning levels that were applied to our subsamples. The amount of work that would be involved in the reprocessing of the archival data with the advanced *sum* cleaning algorithm, is out of scope for this thesis. In addition, the *sum* cleaning hardly improves MAGIC's performance in the energy range above 400 GeV, and hence should not affect our study of the most energetic radiation emitted by the Crab pulsar.

Table D.1 Cleaning method and cleaning levels for the different subsamples

Analysis period	Cleaning method	Cleaning levels [Lvl1/Lvl2]	
		M1	M2
M1 *	Absolute	6/3	–
ST.01.02 (On/Off)	Absolute	6/3	9/4.5
ST.01.02 (Wobble)	Sum	4/3	7/4
≥ST.02.01	Sum	6/3.5	6/3.5

Notes For a description of the cleaning methods, see Sect. 3.4. No distinction was made with respect to the zenith angles in each analysis period. The analysis period ST.01.02 belongs to the time span in which MAGIC operated with two different cameras in MAGIC-I and II, and two different readouts (see Sect. 3.3). For ST.01.02 (Wobble) the levels were optimized by Gianluca Giavitto

© Springer Nature Switzerland AG 2019
D. Carreto Fidalgo, *Revealing the Most Energetic Light from Pulsars
and Their Nebulae*, Springer Theses, https://doi.org/10.1007/978-3-030-24194-0

Pulsar Light Curve Fitting

For the fitting of the phaseogram we define a probability density function (PDF) for the phase ϕ of our events. The PDF includes the two peaks plus a constant in phase accounting for the background, and hence reads:

$$\text{PDF}(\phi \mid P1, \boldsymbol{v_1}, P2, \boldsymbol{v_2}, B) = \frac{P1\, p(\phi, \boldsymbol{v_1}) + P2\, p(\phi, \boldsymbol{v_2}) + B}{P1 + P2 + B}, \qquad (\text{D.1})$$

where $P1$, $P2$ and B are the intensities of the main pulse, the inter-pulse and the background, respectively. $\boldsymbol{v_1}$ and $\boldsymbol{v_2}$ are the model parameters for the peak shape p, which we model with three different normalized functions:

- a Gaussian:

$$p(\phi, \mu, \sigma) = \frac{1}{\sqrt{2\pi}\,\sigma}\, \exp[-\frac{(\phi - \mu)^2}{2\sigma^2}] \qquad (\text{D.2})$$

- an asymmetric Gaussian:

$$p(\phi, \mu, \sigma_1, \sigma_2) = \frac{2}{\sqrt{2\pi}\,(\sigma_1 + \sigma_2)} \begin{cases} \exp[-\frac{(\phi-\mu)^2}{2\sigma_1^2}], & \phi < \mu \\ \exp[-\frac{(\phi-\mu)^2}{2\sigma_2^2}], & \phi > \mu \end{cases} \qquad (\text{D.3})$$

- and a Lorentz function:

$$p(\phi, \mu, \gamma) = \frac{1}{\pi\gamma}\, \frac{\gamma^2}{(\phi - \mu)^2 + \gamma^2}. \qquad (\text{D.4})$$

With the PDF at hand we compute the likelihood function

$$\mathcal{L}\left(P1, \boldsymbol{v_1}, P2, \boldsymbol{v_2}, B \mid \{\phi_i\}_{i=0}^{N}\right) = \prod_{i=0}^{N} \text{PDF}(\phi_i \mid P1, \boldsymbol{v_1}, P2, \boldsymbol{v_2}, B), \qquad (\text{D.5})$$

where N is the total number of events in our light curve. To obtain our model parameters, we minimize the negative logarithmic of $\mathcal{L}$ using the *iminuit* package[6] for python, which is based on the Minuit algorithm by James and Roos [1]. In the case of comparing nested models (for example a symmetric and an asymmetric Gaussian peak shape), we use the *likelihood ratio test* for which the probability distribution of its test statistics can be approximated following [2].

---

[6]iminuit – A Python interface to Minuit. https://github.com/iminuit/iminuit.

Spillover in the >400 GeV Pulsar Light Curve

Due to the finite energy resolution of the instrument, we inevitably will get some spillover of events with a true energy below 400 GeV that cannot be corrected for on an event-by-event basis in our pulsar light curves. Here we try to estimate the percentage of gamma-rays in our >400 GeV light curve, which originate from the inter-pulse of the Crab pulsar and have a true energy below 400 GeV. For this estimation we take our MC events with an estimated energy above 400 GeV and check the percentage of events with a true energy below 400 GeV, after reweighing the distribution to mimic the spectrum of the inter-pulse of the Crab pulsar (that is a power-law spectrum of -3.13). This exercise is done for each subsample of our data set and the resulting percentages are weighted by the number of excess events found in the corresponding subsample (if the number of excess events is negative, we discard the subsample from our estimation). The final number is then the weighted average of the percentages resulting in an estimated spillover of $15.8 + 0.4\%$. The error is estimated by applying a bootstrap method in which we draw the weights from a Gaussian distribution centered on the number of excess events and a standard deviation equal to the error of the excess events. The stated error is then the standard deviation of the obtained distribution of averaged spillovers.

Spectral Points

In Table D.2 we provide the numerical values to the spectral points derived in Fig. 5.8. The spectral points were unfolded by means of the Bertero method for the regularization (see Sect. 3.4), and thus the statistical errors are correlated. The covariance matrices for the two spectra, the main pulse P1 and the inter-pulse P2, respectively, are given on the next page.

The covariance matrix for the spectral points in Fig. 5.8. Matrix D.6 and D.7 belong to the main pulse P1 and the inter-pulse P2, respectively.

$$
\begin{bmatrix}
1.59e^{-22} & 1.46e^{-23} & -1.19e^{-24} & -2.02e^{-25} & -1.19e^{-26} \\
1.46e^{-23} & 1.22e^{-23} & 2.14e^{-24} & -6.95e^{-26} & -2.26e^{-26} \\
-1.19e^{-24} & 2.14e^{-24} & 1.67e^{-24} & 2.18e^{-25} & -1.37e^{-26} \\
-2.02e^{-25} & -6.95e^{-26} & 2.18e^{-25} & 2.11e^{-25} & 2.59e^{-26} \\
-1.19e^{-26} & -2.26e^{-26} & -1.37e^{-26} & 2.59e^{-26} & 2.60e^{-26}
\end{bmatrix}
\tag{D.6}
$$

Table D.2 Fluxes and numerical values for the spectral points in Fig. 5.8

Energy bin (GeV)	Main pulse (P1) $\frac{dN}{dE}$ (cm^{-2} s^{-1} TeV^{-1})	Inter-pulse (P2) $\frac{dN}{dE}$ (cm^{-2} s^{-1} TeV^{-1})
69–108	$(0.70 \pm 0.13) \times 10^{-10}$	$(1.48 \pm 0.14) \times 10^{-10}$
108–167	$(1.17 \pm 0.35) \times 10^{-11}$	$(2.77 \pm 0.38) \times 10^{-11}$
167–259	$(0.31 \pm 0.13) \times 10^{-11}$	$(0.60 \pm 0.14) \times 10^{-11}$
259–402	$(1.02 \pm 0.46) \times 10^{-12}$	$(2.15 \pm 0.49) \times 10^{-12}$
402–623	$(0.24 \pm 0.16) \times 10^{-12}$	$(0.73 \pm 0.18) \times 10^{-12}$
623–965	–	$(1.68 \pm 0.72) \times 10^{-13}$
965–1497	–	$(0.63 \pm 0.29) \times 10^{-13}$
(GeV)	$E^2 \frac{dN}{dE}$ (TeV cm^{-2} s^{-1})	$E^2 \frac{dN}{dE}$ (TeV cm^{-2} s^{-1})
69–108	$(5.00 \pm 0.90) \times 10^{-13}$	$(1.07 \pm 0.10) \times 10^{-12}$
108–167	$(2.02 \pm 0.60) \times 10^{-13}$	$(4.80 \pm 0.65) \times 10^{-13}$
167–259	$(1.28 \pm 0.53) \times 10^{-13}$	$(2.48 \pm 0.57) \times 10^{-13}$
259–402	$(1.01 \pm 0.46) \times 10^{-13}$	$(2.15 \pm 0.49) \times 10^{-13}$
402–623	$(0.56 \pm 0.38) \times 10^{-13}$	$(1.75 \pm 0.44) \times 10^{-13}$
623–965	–	$(0.97 \pm 0.41) \times 10^{-13}$
965–1497	–	$(0.88 \pm 0.41) \times 10^{-13}$

Notes The energy values of the spectral points are chosen such, that the resulting spectral function at these energies is the same as its average in the corresponding energy bin. The resulting spectral functions are given in Table 5.4

$$
\begin{bmatrix}
1.94e^{-22} & 1.66e^{-23} & -1.51e^{-24} & -2.39e^{-25} & -1.08e^{-26} & -4.44e^{-28} & -3.01e^{-28} \\
1.66e^{-23} & 1.41e^{-23} & 2.39e^{-24} & -9.90e^{-26} & -2.34e^{-26} & -1.17e^{-27} & -1.26e^{-28} \\
-1.51e^{-24} & 2.39e^{-24} & 1.89e^{-24} & 2.39e^{-25} & -1.45e^{-26} & -2.60e^{-27} & -2.13e^{-28} \\
-2.39e^{-25} & -9.90e^{-26} & 2.39e^{-25} & 2.42e^{-25} & 2.97e^{-26} & -2.14e^{-27} & -3.65e^{-28} \\
-1.08e^{-26} & -2.34e^{-26} & -1.45e^{-26} & 2.97e^{-26} & 3.36e^{-26} & 4.24e^{-27} & -3.60e^{-28} \\
-4.44e^{-28} & -1.17e^{-27} & -2.60e^{-27} & -2.14e^{-27} & 4.24e^{-27} & 5.20e^{-27} & 6.47e^{-28} \\
-3.01e^{-28} & -1.26e^{-28} & -2.13e^{-28} & -3.65e^{-28} & -3.60e^{-28} & 6.47e^{-28} & 8.55e^{-28}
\end{bmatrix}
\tag{D.7}
$$

References

1. James F, Roos M (1975) Minuit - a system for function minimization and analysis of the parameter errors and correlations. Comput Phys Commun 10(6):343–367. https://doi.org/10.1016/0010-4655(75)90039-9.
2. Wilks SS (1938) The large-sample distribution of the likelihood ratio for testing composite hypotheses. Ann Math Stat 9(1):60–62. https://doi.org/10.1214/aoms/1177732360

Index

© Springer Nature Switzerland AG 2019
D. Carreto Fidalgo, *Revealing the Most Energetic Light from Pulsars and Their Nebulae*, Springer Theses, https://doi.org/10.1007/978-3-030-24194-0

Printed in the United States
By Bookmasters